Thanks Keith,

Please help spread
Compression Thinking

"Doc"

COMPRESSION

Meeting the Challenges of Sustainability Through Vigorous Learning Enterprises

ROBERT W. HALL

CRC Press
Taylor & Francis Group
Boca Raton   London   New York

CRC Press is an imprint of the
Taylor & Francis Group, an **informa** business

A PRODUCTIVITY PRESS BOOK

Productivity Press
Taylor & Francis Group
270 Madison Avenue
New York, NY 10016

© 2010 by Taylor and Francis Group, LLC
Productivity Press is an imprint of Taylor & Francis Group, an Informa business

Library of Congress Cataloging-in-Publication Data

Hall, Robert W., 1937-
    Compression : meeting the challenges of sustainability through vigorous learning enterprises / Robert W. Hall.
        p. cm.
    Includes bibliographical references and index.
    ISBN 978-1-4398-0654-8 (hbk. : alk. paper)
    1. Industrial management--Environmental aspects. 2. Sustainable development--Social aspects. 3. Industries--Environmental aspects. 4. Organizational learning. 5. Organizational sociology. I. Title.

HD30.255.H35 2010
658.4'083--dc22   2009031259

Visit the Taylor & Francis Web site at
http://www.taylorandfrancis.com

and the Productivity Press Web site at
http://www.productivitypress.com

Contents

Acknowledgments

So many people have contributed to the thinking in this book that it is impossible to name them all. It took long enough that I can't remember them all. Several people read substantial chunks of the work and had suggestions, even if only, "I don't understand; please say what you mean." The outcome is not slick writing. Just struggling to clarify original thought took about 10 years.

However, my wife Sandra and Jack Ward both read every word, pointing out inconsistencies and ambiguities and suggesting new directions for research. And both Sandra and my deceased wife Kay put up with a lot while a distracted writer was lost in thought—probing, picking, mulling, and checking.

Ten years ago just trying to discuss this with friends in the Association for Manufacturing Excellence was a bit frustrating. Fortunately, over time they were kind enough to indulge me in presenting bits and pieces of it at meetings. Although at first a good many didn't know what to think about it, that feedback, too, has been valuable. But most of all, the members of the association tolerated me as the editor of their publication, *Target*, for many years. It was a great gift, for it let me see the inside of some of the best-functioning work organizations on the planet. Despite the direction of *Compression* seeming radical, considerable confidence in the proposals came from seeing that many real companies with real people could actually thrive when employing some of the practices described.

Robert W. "Doc" Hall

Introduction

For a long time I was unsure whether this book should be published. Compression is a very different kind of world coming soon—in a year or two to a decade or two. Most of us prefer to ignore the signs of its coming, but the sooner each of us personally puts aside quiet time to think through Compression and its challenges, the better. The subject is too big and fragmented to digest in a single sitting; most of us must mull its implications for years to form the semblance of a cohesive pattern. The writing builds slowly, like my slow conversion from expansionist business thinking to the realization that it is leading us to oblivion. Some of the writing will seem strident, radical—whatever—to those convinced that expansion has no limit, but many points were impossible to make when sugar-coated.

Compression is far too big and dynamic a subject for a book. It can never be completed, and the thinking is systemic. A book format forces presentation to be linear, and can't pack an infinite subject into a finite message. Much is not included. Some "good stuff" is in chapter endnotes (please read them, too), but further evidence and reasoning suggested by many endnotes became too voluminous to print. Please visit the extended footnotes at http://www.productivitypress.com/compression/footnotes.pdf. Please do your own thinking and help add to this story—or better yet, become part of it.

Compression proposes a bottom-up revolution by working organizations and the people in them on whom everyone's quality of life depends. This book is dedicated to these people. They design, build, and maintain buildings, roads, sanitation systems, communications systems, vehicles, electrical power systems, and much more. They extract and refine fuels and minerals. They attempt to prevent illness, but they will care for the sick and heal them when they can. They answer the call in case of fire, accident, and disorder. They grow and prepare food. They teach and mentor the young. They adjudicate disputes. They keep things clean. They research the global environment. They discover how things work, how they can work better, and how we can keep from killing ourselves.

Almost none of us has time to give them much thought, and so they continue to be underappreciated until performance during a crisis reveals just how important they are. They assume true responsibility. And they need a revolution of the systems by which they discharge those responsibilities.

1

Understanding the Challenges of Compression

Hopefully, industrial societies have begun a period of deep, profound revolution in how we think, how we work, and how we regard economics. Despite unrelenting denial, the old order cannot cope with its challenges. A new order, much more vigorous in dealing with the physical problems of the twenty-first century, has to take its place, and soon.

The challenges that must be met by any new approach to organizing work are summarized in Figure 1.1. The four outer circles contain arbitrary headings for a seemingly infinite number of subissues under each one. All are very serious, but in order to deal with them, the biggest challenge is in the central circle: changing us.

Compression, as a word, is used in this book in two ways: First, it refers to the world state we are entering, as shown in Figure 1.1. Second, it refers to what we must do to meet these challenges, to the central circle in that figure. They are interrelated.

Economists accept that humans are not always logical in making economic decisions, but the classical theories that still guide many decisions presume that transactions between "economic men" yield the greatest good for the greatest number. Therefore maximum profit must be the primary objective of any economic organization.

Capitalism has never lacked critics. Karl Marx's main beef was that it is unfair. He was right; it is. No other system is fair either, and some folks have a low whine threshold. But when unfairness exceeds the tolerance of a critical mass, people upchuck an old system, as in Russia in 1917, and then again in 1989. But today this business system has a more serious flaw: incompetence in dealing with the challenges of Compression.

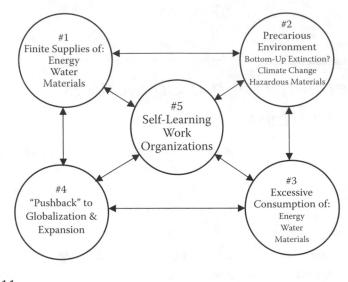

FIGURE 1.1
Twenty-first century challenges of Compression.

Any new system emerging from it must be more flexible than big-scale capitalism, have greater proficiency inventing and deploying new ideas, and deal with the challenges in Figure 1.1. That's a tall order. In addition, it has to quickly resolve fairness disputes before they undermine the dedication of those working for the good of others.

Think of this as a big leap in enlightened, collective capitalism. Despite its flaws, the capitalist profit motive tempered with service motives usually performed well, delivering productivity, quality, service, and technological advancement. Real stinkers went broke. Centralized socialist objectives such as maximum employment and output targets were too inflexible. Communists were even worse at conserving resources and preserving the environment. But both systems were designed for economic expansion, consuming more and more resources.

To deal with Compression, twenty-first century work organizations crucial to human survival need a new mission: assure survival of life and promote quality of life using processes that work to perfection with self-correcting, self-learning systems. No use of excess resources. No wasted energy. No toxic releases. Quality over quantity, always.

Quantifying this sharpens the impact: flawless performance providing an industrial society quality of life to as many as possible while consuming less than half of the energy and virgin materials used in the year 2000, and with zero toxic releases. Our crucial work organizations need missions that focus on some aspect of this challenge without fracturing along the fault line of endless twentieth-century wars and disputes. Economic emphasis must shift from how to divvy up what we can get, to what we must *do*.

No one has the time to track fast-moving research in scientific journals, business news, political news, and tech manuals. Without seeing patterns, one only grasps disjoint fragments. Even this tome can only present a few points in support of each Malthusian Challenge (the most famous of all doomsday forecasters).[1]

CHALLENGE #1: RESOURCE SHORTAGES

Four commodity resources are vital to an industrial society quality of life: energy, materials, water, and air. In any physical industrial system, they interrelate; they're not separate cost items. Shortages of fossil fuel energy are imminent for two reasons: lack of long-term supply and environmental damage from burning them, but energy is needed on a big scale for even a parsimonious form of industrial society. At present all alternative energy sources have the disadvantage of a lower energy yield than fossil fuels. Without a novel breakthrough, a future economy cannot be based on so-called cheap energy. Water is now critically short in many places. Without major improvements in usage efficiency, these will foreclose an industrial society quality of life to many people. Virgin sources of common metals such as iron and copper are ample, but continuing to increase production is problematic.[2] Aluminum is an extreme example; recycling it uses only 5 percent of the energy needed to refine it from ore. Petroleum feedstock for polymers is not inexhaustible; innovative chemistry is needed. However, many materials are reusable, so their availability depends on the energy required to reuse them.

Fossil Energy

Despite its economic importance, Hubbert's Peak has barely entered public awareness. The term refers to the time at which an oil field, or a group of them, reaches maximum output and thereafter declines. Peak does not mean suddenly running out. It means that ever-growing demand cannot be supplied.[3] Petroleum is not only burnt for energy, but a small percentage of it is chemical feedstock converted into plastics and other materials.[4]

Oil fields do not suddenly go dry. They slowly decline. Wringing out all the oil takes more and more energy, beginning with secondary recovery: injecting water or steam into a field to force more oil into the extraction pipes, or horizontal drilling to expand the area from which pipes suck up oil. Tertiary recovery uses detergents and other methods to free oil trapped in rock. Once into secondary and tertiary recovery, the pumping rate of a field slowly declines, like squeezing the last drop of juice from a lemon. More work—energy—is required to get less out. That is, the energy yield of a field declines. Overly aggressive recovery forces oil further into the rock instead of toward extraction.

The energy yield of the first Texas oil gushers may have been 100:1. The Canadian tar sands are more like 3:1, perhaps closer to 2:1.[5] Physical energy economics are affected by dollar demand only if it inspires higher energy-yield methods for the total processes from discovery through actual burning for use.

Increasing market demand doesn't put more oil—or coal or natural gas—in the ground; so in recent years, interest in predicting Hubbert's Peak for fossil fuels, especially petroleum, has become more intense. The data used for this are imprecise. The quality and size of oil reserves, extraction rates, and future market demand are all subject to variance—and to fuzzing for PR and negotiating purposes. However, satellite imagery is sufficiently advanced to make it unlikely that any untapped fields are close to Saudi Arabia's Ghawar field in size.[6] Pessimists think peak oil is here; optimists put it at 2020 or beyond; but nigh enough for serious preparation.[7] A natural gas peak is even more uncertain.[8] But political disruption of existing supplies is the most imminent threat.[9]

Energy yield is crucial in evaluating alternative biomass fuels. Replacing motor fuel with them would consume all U.S. crop ground. The most optimistic estimates project that no more than 30 percent of the current volume of motor fuel used could be grown.[10]

All presently known alternative energy sources have low energy yield or pollution, or both. Oxidation of every fuel molecule reduces the pollutants (other than CO_2) to be scrubbed from exhaust; scrubbing cuts energy yield. Coal has less energy density and a lower energy yield than oil. Thus, clean coal's sequestering of CO_2 is possible, but at a lower energy yield, and is untested at commercial scale.[11] Nuclear fission has an acceptable 4:1 energy yield, but well-protested safety and environmental issues; and uranium ores appear to be decreasing in concentration. Nuclear fusion is decades from commercialization, if ever. Photovoltaic cells have promise, but for viable energy yield, need better solar conversion efficiency with lower energy production processes.[12] Biomass fuels' yield is about 3:1 at best (and below 1.0 in inefficient ethanol plants); one can't get more energy out of a source than nature put into it. When growing biomass for fuel robs ground from food crops, global food shortages drive up food costs.[13] Growing food also takes energy. High crop yields now depend on nitrogen fertilizer made from natural gas, not infinite in supply. Using a fuel to make fertilizer detracts from the energy yield of biomass, whether used for food or fuel.[14] Food adequacy is related to every quality-of-life challenge of Compression, so that diverting energy from growing food to growing fuel is only one issue.[15]

Hydrogen is an energy carrier, not a natural fuel. If made in volume today, the feedstock is natural gas, but converting it to hydrogen cuts its energy yield.[16] In addition, only 10 to 20 percent of the fuel energy coming from a wellhead turns the wheels of a vehicle. Every part of our industrial system has huge energy losses—and huge opportunities to reduce them. The real test of an energy source is whether, for example, energy from photovoltaic cells can be used for the entire process of making more photovoltaic cells, from raw material to installation, with plenty of energy left over for other uses.

Energy is vital. Everything, including people and computers, runs on it. Thermodynamically, the agricultural age ran on food energy and wood fuel. People obtained bodily energy from food, and burned it using hand tools. Animal power came from food, too; preindustrial farms used about a quarter of the ground raising feed for work animals. Today, the engines and electric motors that power an industrial society, including its farms, mostly run on energy originating from oil, gas, or coal.

The industrial revolution started with hydropower water wheels, but its expansion took off on coal in the nineteenth century, then oil, then nuclear fission. Technology to capture and use ever-increasing sources of energy on a big scale (for all uses, including military) succinctly explains why a system based on European colonialism became dominant.

Free energy available for other uses is that in excess of that required for system maintenance. In the case of living organisms, free energy is that in excess of basal metabolism, the minimum energy necessary to maintain life. Wild animals use a great deal of their energy above basal metabolism finding food. Factor that in, and free energy to spend above bodily maintenance is more limited. The more dispersed their food, the lower their net free energy. Likewise, the energy yield from any fuel is cut if it is widely dispersed. Energy spent gathering energy saps potential human society free energy, which is why an overall declining energy yield from sources tapped should ring alarms.

Joseph Tainter, an archaeologist, uses energy efficiency to explain the rise and fall of ancient empires whose economies were not masked by anything like our mass transaction systems. Empires like the Chacoan in the American Southwest appeared to expand by conquest, exacting tribute, labor, and food from their neighbors. This gave them more free energy to squander on elaborate housing, games, and—inevitably—monuments to their own glory. In effect, they taxed the energy from conquered land. Some empires fell to stronger superpowers, epidemics, natural disasters, and exhausting their ecology. [17] But those like the Roman Empire withered away when they could not sustain the free energy to defend large territories and long supply lines. In any case, an ascending empire imported energy by subjugating others. An empire declined when it lost the *mojo* to sustain the energy supporting the style to which it had become accustomed.[18] Peter Turchin calls this *mojo* an empire's "high capacity for collective action."[19]

Translating this analysis of the ancients into a modern context, if the developed world depends on energy, but its energy yield from the aggregate of all energy sources is dropping, one of two consequences ensue: clean, concentrated, higher-yield energy sources are discovered. Or it has to stop wasting its free energy and learn how to accomplish the same objectives using less energy. That is, it enters Compression.

Fresh Water

Humans have a long history of both sharing water and fighting over it. Water is symbiotic with life. Living beings are water bags full of biochemistry. Despite this, the scientific understanding of the physics and chemistry of water has big gaps. Little about water is as simple as it seems, not even where it might have come from.[20]

Water is used for everything, from power generation to sewage treatment, irrigation, and cooking. Humans value water as a commodity, and even as a basic right, but usable water is not plentiful. About 97 percent of global water is saline; 3 percent, fresh. Of this fresh water, about 30 percent is underground, theoretically possible to access, but only 0.006 percent is in rivers, the main human source.[21] Humans have to either go to this water or bring it to them. Some people expend hours and lots of calories carrying water every day, while it is pumped to many of us. About 2 or 3 percent of all fuel energy (roughly 20 percent of the world's electrical energy) pumps water. Ninety percent of the lifetime cost of a pump is for energy.[22] About 70 percent of all fresh water use is for agriculture. Opportunities to improve this are huge.

In much of the world, water scarcity has been the norm for millennia. Privatization of water systems versus regarding them as a common resource is an issue. Because of its history, even in the United States, the bastion of private enterprise, private management of water raises more hackles than in Europe. In the nineteenth century, many American cities bought out private owners, exploiting easy sources of water, but served only customers able to pay. The wealthy got water; the poor didn't; and water quality was often far short of Six Sigma. Such a legacy dies hard after the specifics are long forgotten.

Where water has always been scarce, wells and other water sources have long been regarded as open for everyone to use. Private control of a water system is a strange idea. But as populations expanded, old communal systems could not provide water in the quantity and quality desired. Systems had to tap bigger sources using electric pumps, expanding territorial coverage and water use per capita. If governments can't raise public debt for this, turning water over to private companies is politically attractive. It substitutes the private company's capital for public debt, and shifts its taxes to the company's water fees—along with the backlash when people are expected to pay. In most cases an emotional explosion is rooted in both dire poverty (no cash) and in cultural legacies for managing common resources, reasoning that if water (and oil and gas) are common resources that no one owns, why should anyone buy or sell them? Because of the cultural clash, the sides in these disputes uncomprehendingly talk past each other.[23]

Where water became plentiful, it became easy to overuse. We understand too little, for instance, about how underground aquifers recharge.

Before electric submersible pumps it was almost impossible to overpump an aquifer, so we are learning the hard way about this.[24] Flow rates were obviously limited from aquifers tapped by Middle Eastern *qanats*, horizontal boreholes into distant mountains. A few date back 3000 years. Common usage of water or any other resource has been called the "doctrine of usufruct": anyone can use what they need, but no more, and must not ruin the source for others' use.[25] In shortage conditions, it was obvious that everyone's life depended on such customs even if they were mortal enemies. Usufruct was supported by religious beliefs, and forms of water worship lingered in Europe until about 1000 A.D.[26]

However, the capitalist doctrine of self-interest saps the instinct for sharing commons, a weakness often called the Tragedy of the Commons.[27] For example, private-interest herdsmen could keep their independent cash outlays down by each using a common meadow until it was collectively nibbled to the nubs. Preindustrial people may be less inclined to this, but inclination is insufficient. Several ancient civilizations are thought to have overused their environment, particularly when hit by drought; and cultures that have usufruct today are not staving off the effects of severe shortages. Look at the subtexts in the news, and water shortages are a factor in calamities such as Darfur in Sudan and its surrounding regions.

Slow declines do not make the news. Clash and conflict do. But water shortages are severe in many areas. Eight major rivers are tapped out.[28] And still, damming of rivers continues. The Three Gorges Dam on the Yangtze in China is either highly lauded or cursed, depending on the effects from it one chooses to see. China has commissioned a $50 billion canal project to divert water from the Yangtze to cities like Beijing.[29] More Chinese projects are slowing the flow of the Mekong through countries downstream.[30] There, people fear that the Chinese are stealing their water, or just as bad, ruining it for their use, reducing its natural ability to cleanse pollutants and grow food. A well-known constant battle against these effects is in the Nile Delta as a result of the Aswan High Dam. Fertilizing and irrigating ground formerly silted by flood ruins cropland. Those issues remain manageable; not so debacles like the Aral Sea. Possibly the most contaminated site on earth is the radiation-laden old Soviet Union testing area north of the Ural Mountains, Novaya Zemlya.[31] A big factor in preserving the quantity of fresh water available is also preserving its quality. The U.S. Geological Survey sees the prevention of chemical contamination and eutrophication (oxygen depletion) as major challenges.[32] Contamination of ocean water also threatens the biosphere in multiple ways.

If we have the will, we can do much better. Modern municipal water systems have working organizations that can begin to deal with their problems, and some have. For example, both Israel and Singapore are on the ragged edge of water self-sufficiency—already in Compression. Their practices start to meet these challenges. In both countries, unaccounted water losses are well under 10 percent, very good by global water system standards. Both recycle and reuse water. Both are desalinating salt water.[33] Singapore plans to capture run-off from a large portion of the city, purifying it by reverse osmosis to ease both sporadic flash floods and water shortages at the same time. (Singapore is surrounded by salt water, not fresh water.) In both systems, fee structures discourage high usage—thus compressing the water footprint at the tap. Irrigation of Israeli crops is a major use in that system, so engineers laser-level irrigated fields and devise slow-drip, low-evaporation systems, trying to feed every drop to a plant. These measures presage the future of water systems in many industrial societies.

CHALLENGE #2: A PRECARIOUS ENVIRONMENT

The environment can be damaged in so many ways that not even a committed environmental scientist can track them all. Many of them seem related to global warming, which draws attention because it implies that a big reduction in CO_2 emissions from burning fossil fuels could ameliorate it. This seemed so threatening to some existing economic interests that at least one disinformation campaign attempted to fuzz public understanding of it.[34] However, the term "global warming" has become a catchphrase for a changing balance in the global biosphere. A projected temperature rise is only one indicator of it. A shifting CO_2/O_2 cycle affects the biosphere balance in many ways. While CO_2 is the most voluminous of greenhouse gases, it's not the most potent of known catalysts that could tip the balance of the biosphere in a direction unfavorable to human well-being, or even survival. The present biosphere balance favoring humans is sometimes called the Goldilocks Zone. The objective is to stay in it.

Given the multitude of ways the biosphere can be unbalanced, the key objective is to maintain it so that six to eight billion humans can live with a quality of life somewhat like that in industrial societies today, if not better. Scientific papers presenting evidence of threats and problems are generally detailed and devoid of context for a general audience. And many problems are so interconnected that connecting the dots in even a superficial way takes more than casual attention. Most likely, no single person can grasp the full picture.

Precarious describes big swaths of the global environment. Anticipating changes by linear projections is insufficient because slow degradation may be disregarded until an ecological system under stress passes a tipping point and suddenly breaks down. Media coverage of complex threats tends to spotlight predictions of catastrophe, but precursors to a catastrophic tipping point may not be noticed at all, and if so, given little media play. To project the future of our biosphere, science is forced to model phenomena never before seen, basing assumptions on related phenomena and evidence. In classical science, experiments check the validity of explanatory hypotheses; but if a hypothesis is more like "a rise in temperature might make a bomb you are sitting on explode," the actual experiment would destroy its observers and all others around it. Few observers are eager to conduct such experiments.

Agreement projecting complex threats is difficult enough to forge among insiders to the science and the models. Media attention is drawn to worst-case possibilities, sometimes in simplified form. Airing the insiders' drawn-out arguments about model assumptions or conclusions in the media does not generate public confidence in a consensus on the evidence if one is ever reached.

An example is the potential for a volcanic eruption of Cumbre Vieja volcano on La Palma in the Canary Islands off the coast of Africa, investigated by Prof. William McGuire, director of the Benfield Hazard Research Center, University College London, and students. The center is a research source for insurance companies on both long-term and short-term natural hazards, and so monitors volcanoes, earthquakes, and tsunamis. McGuire specializes in volcanoes and rare mega-catastrophes. In the 1990s, he became interested in the possibilities of quake-induced tsunamis in the Atlantic Ocean. All the monster tsunamis in recorded history have been in the Pacific. One site studied was Cumbre Vieja, a volcanic peak on La Palma in the Canary Islands.

The study concluded that it was possible for a major quake to cause a flank collapse of a huge slab of rock, up to 500 cubic kilometers hitting the ocean at 100 meters per second. That would set off a tsunami bigger than any historically seen. Possibilities depend on how a geologist reads the faults in the rock face, comparing them with rock falls of comparable size that seem to have occurred several times over the past million years or so, but that was not emphasized in a study looking for the worst case. A follow-on study modeled the tsunami that could occur from a worst-case flank failure. It concluded that this energy would propel a tsunami across the Atlantic in about nine hours, hitting the coast of the United States with a succession of waves 20 to 25 meters high, tall enough to wipe out Miami, for instance. Half as much rock hitting the water takes the wave size down to about a fourth that height—devastating, but not a total wipe-out.[35]

Newsmakers seeking adrenaline-juicing material had a field day with this, often neglecting the context and assumptions, and sometimes distorting the basic scenario. One report even said that water gushing from Cumbre Vieja would initiate a tsunami. Many claimed that a disaster was imminent. A group called the Tsunami Society attempted to debunk the claim.[36] Natives of La Palma were dismayed; the stories scared away tourists. Google Cumbre Vieja, and one finds an eight-year history of hype and counter-hype.

Reading the original papers suggests that the authors sought the mechanisms of a worst-case possibility without being explicit about their assumptions. Yes, mega-tsunamis very likely occurred over a stretch of a million years or so, but they're unlikely in a window the size of a century, and even less within a decade. But the studies triggered a tsunami of alarms still washing about, while geophysicists continue to explore the evidence—and models—for mega-disasters all around the world. McGuire is one of them. The *2007 Hazard and Risk Science Review* from his center says nothing special bout Cumbre Vieja.[37]

Cataclysmic meteorite hits or perhaps mega-seismic events are hypothesized to have been climate-changers in prehistoric times. Whether lesser such events have had significant influence in historic times is more conjectural. Ash from major eruptions such as Krakatoa and Pinatubo is known to have depressed global temperatures for years until the ash died down. Such events are a part of ecological history.

Public discussion of such subjects tends to diverge from scientific objectivity for at least three of the following reasons:

- Researchers fail to frame the intent, scope, assumptions, and limitations of their studies.
- Reporters don't know this context, and if they do, can't frame it in a short, readable story. Consequently, they report fragmented threads,

sometimes with errors repeated so many times that they are assumed to be true.[38]

- The public does not comprehend the context, scientific evidence, scientific reasoning, or debates about interpreting data.

But the crucial issue is whether human-influenced disruption is tipping ecological balances, regionally and globally, so as to endanger the life and health of the human species. That can happen in many ways, some of which may be unsuspected by anyone.

To make this point, only three of many examples of current "concerns" are chosen: honeybees, the ozone hole, and oceanic phytoplankton.

Honeybees

Bred by beekeepers for centuries, honeybees are no longer plentiful—or free. Wild honeybees, escaped descendents of domestic bees bred for centuries, are now almost gone in many parts of the United States, choked out by mite infestations, especially Varroa mites, which infect domestic beehives, too. Honeybees are vital to pollinating many fruits and nuts common in the American diet, so for about 30 years, farmers have been paying beekeepers to manage the infestations and truck domestic hives to the farm. So far, efforts to breed a mite-resistant honeybee have not succeeded.[39]

On top of this, a record outbreak of colony collapse disorder (CCD) afflicted honeybee hives in the winter of 2006–2007. About 23 percent of all honeybees in domestic hives simply disappeared. Media and political attention suddenly focused on this previously quiet plight. A CCD Working Group of researchers formed to hunt down the root cause. Trucked commercial bees appeared more susceptible than those in stationary hives. Use of pesticides to combat the mites and other pathogens might contribute. A virus from Israel (IAPV), discovered in 2004, is generally present in bees with CCD, but definitive causality has not been proven. The honeybee genome has been mapped. There's a project to track the evolution of viruses. So far, evidence points in multiple directions, so many researchers suspect that a combination of stresses taxes honeybees' immune systems beyond their limit. If so, we need to breed either a more resilient honeybee or learn not to stress them to their breaking points.[40]

Honeybees' duress illustrates a more general question: are other key species, like pollinators in the wild, having sustainability problems that we won't notice until the consequences set off alarms?[41] Frog and bat disappearance likewise disquiets the news. What happens if they go?[42] This is the serious side of biodiversity loss. Are we retaining big enough genetic pools in nature to prevent disasters that genetic engineering cannot offset? Pessimists are engaged in such things as maintaining ancient seed banks. We may need every codon of DNA history we can get to modify crops to ward off climate change, disease, competing weeds, and predator population explosions.[43]

Freon and the Ozone Hole

No longer growing rapidly, the ozone hole is out of the news, but still there. In the southern hemisphere's spring each year, high-altitude ozone thins out, letting an increased amount of UV-B light strike the earth around the South Pole. That's the ozone hole. A similar but smaller phenomenon has been detected over the North Pole. Until fingered as a catalyst increasing the destruction of high-altitude ozone, DuPont's Freon was considered to be the ideal safe, effective refrigerant, and it had many other uses: aerosol can propellant, foaming plastics, and cleaning printed circuit boards.

Discovery of the ozone hole illustrates the complexity of our precarious environment. It's a real phenomenon, not a computer model projection, but so abstract from daily experience in an industrial society that full explanation is impossible in media sound bites (or in a few paragraphs). Once loose in the atmosphere, Freon takes a long time to break down, so the hole won't revert to normal size for 40 to 60 years.

Curious British scientists began studying both the seasonality and chemistry of the ozone hole in 1957. Its annual size followed a regular pattern until the mid-1970s, when it began to grow. Chlorine released by fluorocarbons, like Freon, attaches to droplets in stratospheric clouds that form only in the cold Antarctic winter. Chlorine is a catalyst intensifying the normal breakdown of high-altitude ozone. Although Freon is only one fluorocarbon chemical, it releases chlorine readily and was present in the largest concentration.

By 1975, the ozone layer seemed to be growing thinner, but the trend in the data was not at all conclusive. The scientists began to examine the hole more closely, mapping its size and pinning down the chemistry. Were they seeing a normal statistical variation? Missing some measurement errors? A blue-ribbon team of statisticians reviewed the data. DuPont, the primary producer of halocarbons, including Freon, contributed to the team. By 1984, both the trend data and its consequences clearly pointed toward danger.[44] The team rang a global alarm. It was greeted with great disbelief.

But more and more data coming in made the case open-and-shut, at least to scientists and statisticians. In 1987, amid grumbling, the Montreal Protocol initiated the global phase-out of Freon. Companies had to find substitute refrigerants, aerosol propellants, and cleaning agents. By 1992, large-scale use of Freon had ended, although it can still be found on the black market.

But the drama of the ozone hole may not be over. Freon is not the only catalyst destroying atmospheric ozone. Among several others to watch is HFC-134a, a substitute for Freon. Its atmospheric concentration is rising rapidly. It breaks down in the atmosphere faster than Freon, so it is considered less dangerous; but if we pump it into the atmosphere in unrestrained quantities, what then?

Why is the ozone hole dangerous? Ozone absorbs UV-B light on its way to the earth's surface. Excessive UV-B damages *all* forms of life. Exactly how catastrophic this could be is unknown, but UV-B radiating over the earth's surface well beyond the Antarctic could be devastating. UV-B stresses all plants and animals, including oceanic plankton, essential for all life. By comparison, human skin cancer is a trivial problem.

Oceanic Plankton

Phytoplankton and zooplankton exist at the bottom of the ocean food chain.[45] Seriously depleting organisms at the bottom of the food chain is no trivial matter. Besides cutting the food available for the ocean fish catch, photosynthesis by these little critters accounts for nearly 50 percent of all global oxygen production.[46] By contrast, tropical forests, which get a lot of attention, contribute only 17 percent to global oxygen production. Plankton are not as catchy a theme for Earth Day as tropical trees.

Substantive depletion of plankton could tip the amount of oxygen in the atmosphere below 20.9 percent, where it has been for millions of years. Low oxygen would affect the food supply all the way up the food chain, changing all life on land as well as sea. That's bottom-up extinction, and humans would be vulnerable. Our big brains need a rich diet and plentiful oxygen. Cold-blooded reptiles or creatures that suspend animation tolerate diminished food and oxygen better. Because bottom-up extinction is not easy to contemplate without going mad, scientists seldom mention it. However, that unspeakable dread was one motivation to get control of the ozone hole.

Top-down extinction is easier to understand. It is a die-off of large species, like bison in the American West—cute animals we can relate to. Big ones are easy to see, but they do more than add to the scenery. For instance, big animals tramp up soil so that more seed takes root. If one dies off, regional ecological balance is not the same, but it can rebalance to patch over a top-down extinction, and life, even if diminished, goes on. However, a bottom-up extinction diminishes the energy trains running to everything above it. In that case, an old ecology substantially terminates, and a new one starts to evolve nearly from scratch. (The late Steven Gould called this a punctuated equilibrium in the evolution of life.[47]) Such an event is unlikely to favor continuance of human life as we know it.

Unfortunately, environmental worriers' concern for plankton productivity has reasons other than UV-B radiation from an ozone hole. Plankton release oxygen to the air after separating it from hydrates absorbed in the ocean. They concentrate in temperate zones, and nearer land, where nutrients "upwell." They're less abundant in tropical waters. Satellite photos show northern plankton blooming in the spring, like plants greening up on land. Plankton had been thought to be fairly robust to ocean warming and acidification, but a 2002 study reported that spring blooms of phytoplankton off Northern Pacific coasts have decreased about 30 percent; total ocean primary plankton productivity was estimated to have decreased by 6 percent.[48] It's too early to tell how serious this might be. Neither the carbon-oxygen cycle in the ocean nor the consequences of diminished oxygen on all life are well understood. For example, huge masses of anaerobic organisms (archaea) below the sea floor generate CO_2; how that affects plankton and the oceanic carbon cycle is still mostly guesswork.

Not even basic photosynthesis is well enough understood to make confident predictions or offer prescriptions. However, lack of light unquestionably suppresses it. Estimates by a variety of methods have determined that in the last half of the twentieth century, the average amount of sunlight hitting the earth's photic zone has diminished by 10 to 20 percent—during a cyclic upswing in both solar intensity and global temperature.[49] That's why simplistic quick fixes, like shooting an artificial cloud into the upper atmosphere, are dubious. Scientists living biosphere science, and seeing how interconnected it is, sense that a miracle fix could easily improve things worse.[50]

Other trouble spots could also turn out to have ozone-hole effects. For example, coral reefs in the tropics are bleaching. (A dying reef turns white—bleaches.) The base of the reefs is composed of zooxanthellae, a form of algae. This die-off is related to ocean acidification, possibly more threatening to life than atmospheric temperature rises, storm patterns, and rising sea levels—all much easier to portray in media accounts.[51]

So what? Limited evidence suggests that bleached reefs may reconstitute themselves to become more robust to acidification. But if all ocean reefs disappeared, with them would go much of the bottom of the tropical marine food chain. All other things remaining equal, some of the food chain to humans would be gone, and not all humans might survive that. The biosphere has reconstituted itself in modified form many times over in eons past. The issue is whether an industrial-level quality of life can survive a major ecological rebalance. Our ignorance about this is a problem worse than fossil fuel limits, and we don't know how to learn very fast.

Unfortunately, our precarious environment now has more red flags flying than when the ozone hole began to open in the 1970s. No one really knows what unchecked UV-B radiation might have done to all life. Likewise, no one knows the long-term cumulative effects of widespread use of most of the 60,000 or so cataloged industrial chemicals. Having to take action despite many unknowns exacerbates "wicked problems," which is the term for social messes complicated by our own ignorance and seemingly intractable conflicts.[52] We inject ourselves into each problem by arguing whether one exists. Issues boil down to how much risk to take with unknowns. An old adage is, "What you don't know won't kill you." Were that true, no one would have ever walked into a booby trap.

CHALLENGE #3: EXCESSIVE CONSUMPTION

It seems possible to enjoy an industrial society quality life using a fraction of the energy and materials now consumed. Although far from original, that idea is a counterintuitive hypothesis of this book. Using fewer resources—doing much more with much less—will substantially ameliorate the problems of Challenges #1 and #2 despite a high degree of ignorance about them. However, the bias of twentieth-century business and economics is toward growth and expansion—using more and more, often to no purpose other than something to do. Viewing the economy as physical activity, and using old-fashioned terminology, economic man is a profligate wastrel. A couple of centuries ago, before industrial-grade expansion, we did not command the fuel energy to create many of the things that each individual in an industrial society now consumes with barely a thought.

But economic man in an industrial economy also enjoyed a much better quality of life based on almost any indicator. Longevity is one. Between 1850 and 2004, the life expectancy of a white female in the United States went from 40.5 years to 80.8 years. And life expectancy in the United States is not the highest in the world, and barely higher than in Cuba.[53] Quality of life is in the opinion of those living it, and how long one lives is not strictly a function of GNP per capita.

In 200 years, global population has grown at least six times (living a lot longer helped balloon it). Had per capita consumption remained the same, the footprint of human resource consumption would be about a dozen times greater. But per capita consumption of physical resources also multiplied many times. How many times can only be quantified by guesswork, but one rough indicator suggests a startling rise in American consumption in just 45 years. Between 1960 and 2005, American municipal solid waste increased from about 90 million tons to 245 million tons, and per capita solid waste rose from 2.7 pounds per day to 4.4 pounds per day. Per capita daily waste began to tail off beginning around 2000, but total waste tonnage is still rising.[54] Population growth is just one aspect of the larger problem: rapid expansion of resource consumption.

Human population growth is often blamed for excess consumption and environmental problems. Malthusian arguments have raged off and on for 200 years. Expansionists discount them because so far earth keeps carrying more and more people. Obviously, earth's carrying capacity is not infinite, but it's a limit nobody can pinpoint. All kinds of models have been developed to do this, and those most quoted extrapolate estimates of ecological footprints.[55] But plenty of evidence suggests that humans are overconsuming resources in two ways. One is the inability of an ecosystem to support so much as a local population, as in sub-Saharan Africa. The other is excessive resource use by industrial societies, which draw many resources from distant ecosystems.

Market fundamentalists assume that increasing wealth will automatically result in population growth leveling off at earth's capacity limit because it is statistically associated with decreasing birth rates, so that population self-regulates. Therefore the current system will confer a good quality of life on a large population without making conscious changes in either the system or in human behavior. However, that overlooks per capita consumption, which has risen rapidly in industrial societies, and which varies widely, both within economies and between them. Experience with market economies so far is that some folks chew up massive resources; others live on practically nothing. That imbalance is assumed correctable by trickle-down, a rising tide lifting all boats. This logic path doubles back on itself to rationalize continued expansion while also ignoring its own well-known market psychology: overshoot-and-collapse. Optimistic traders discount omens of doom, assuming that the market will invent fixes for any problem.

Discounting human tragedy as mere social Darwinism is not a view exclusive to market fundamentalists. James Lovelock, who first proposed the Gaia hypothesis (the global biosphere is a single interconnected system) thinks that earth's carrying capacity might be only half a billion to a billion people. He offhandedly proposed decreasing its population to that level by 2100. Any wars or pestilences by which this might occur would be nature's way of enforcing its rules.[56] Mad Max scenarios are not hard to project.

To be successful in achieving a longer, better quality of life, we have to think through the physical changes that would prevail were we actually doing that, and then work backward to devise how to get there. One example of such a change would be demographic profiles almost flat in age, unlike prior human experience, but now emerging in mature industrial economies. However, many social conventions, like retirement and health systems, still presume a big base of youth supporting a few oldsters. Almost no country has begun to adjust to a flat demographic age profile, including China, whose birth-averse policies are intended to let its population level off without catastrophe. A third or so of everyone's life would be outside the workforce as we now define it, either as youth, or as fully retired elderly. Every two working people would have to support one nonworker in some way—as a dependent, through high social system premiums, or both. But the implications of such issues run far deeper than public policy by the current system. If an economy does *not* exist primarily to expand production and consumption, that changes the very definition of work, how people learn to work, and many commonsense assumptions about life and its economics. Work becomes adult professional responsibility for activity critical to life. But unless we change our basic premises from expansion to Compression, we can't think our way through this.

To begin doing this, think of economics as physical things and activities, not as financial representations of them. Much about the physical economy you can see directly; no monetized aggregate statistics are needed. The physical economy acquires, processes, consumes, and discards huge quantities of materials, energy, water, and air. Indeed, converting nature into human-fashioned material is a major factor in creating what we call financial wealth. Real estate development is a prime example. Wealth is only how we monetize such activity—quantify a human value for it. By both monetary and physical measures, the United States has long been the world's undisputed consumption champion; and the top seven industrial economies together—about 10 percent of the world's population—account for over half of all world consumption expenditures.

The American way of consumption cannot extend to the entire world. Looking at physical consumption reveals why. American suburban sprawl depends on automobiles. The United States still has about a third of the total world vehicle registrations (not counting motorcycles, lawn mowers, boats, and tractors). If India and China had the same vehicle per capita ratio as the United States, the world population of almost a billion vehicles would nearly triple. With anything like present technology, the most optimistic of fuel forecasters does not know wherefrom the fuel for this many vehicles would come. At the present rate of expansion, the world vehicle population will surpass a billion registered cars and trucks around 2010.[57]

Annual unit sales drive the automotive industry. Few of its leaders pause to think of the consequences of the cumulative numbers pile-up—until stuck in a traffic jam. About 3 percent of all motor fuel is burned in traffic jams. In urban areas, another percent or two is used just looking for parking places. If all American cars (not including trucks) were parked in a Walmart style parking lot, the thing would cover an area approximately the size of Lake Ontario.[58] Because cars take up space, development of everything else to depend on them pushed American physical economic activity to adopt an automotive layout. Exurbs keep expanding; commuters measure distance by drive time. Except for a few old compact cities like New York City, communities laid out for walking are now so rare that they are news curiosities. In most communities, one has to drive to do almost anything. American physical economic development assumed cheap energy with few fallback alternatives.

During the twentieth century, the number of units annually added to housing stock ballooned 26 times.[59] People acquiring more stuff needed space to keep it. After World War II, the average new house expanded from about 1500 square feet to 2320 square feet, with at least a two-car garage attached.[60] Americans who prefer to drive their house can now obtain a motor home with a built-in garage for small vehicles.[61]

If they had the means, nineteenth-century American travelers stayed in hotels or in friends' houses. Now, interstate highway exchanges with a half-dozen motels are common. The national stock of homes away from home ballooned too.

The average American burns 126 pounds (about an adult body weight) of petroleum per week for motor fuel plus other uses. Considering all energy consumed from all sources, each American burns the gasoline *equivalent* of 350 pounds a week (about 50 gallons). Americans swig, on average, 50 gallons of soft drinks each year, more than a standard 12-ounce can per day.[62] The calories accumulated in American obesity have been estimated as sufficient to feed Afghanistan for a year.[63] Figure 1.2 illustrates how both total American consumption and per capita consumption burgeoned during the twentieth century.

Between 1990 and 2000, mortgage debt rose 81 percent while the number of housing units rose only 13 percent; the size and value of homes increased, and consumers consolidated debt in tax-deductible second mortgages, which is believed to have slowed the rise in credit card debt. This dated data stops before the mortgage debt bubble began in about 2003. According to one data series, household mortgage doubled again between 2000 and 2006, but the picture you would have seen during this run-up depended on the data series tracked.[64]

World-class consumption was promoted with world-class persuasion. Saturation marketing became part of the culture. Over half of all world advertising spending is in the United States, $243 billion in 2000, exceeding the GNP of all but 20 other countries. Every day, $2.50 worth of ads tries to penetrate each American's psyche, like a swarm of sperm all going for a buying-decision egg. Although that's more than the daily income of about half the world's population, it's dwarfed by spending on other forms of persuasion built into the system.[65] From Tupperware™ to software, high-markup products burn half or more of their revenue in marketing. And after decades of

Year	Percent DPI Spent on Food	Eat-Out Percent	Gal. of Cola Consumed	Million BTU per Yr.	KWH/yr. Electricity	Consumption $ per Capita (in 1996 $)	Consumer Debt in billions of $
1930	24.7	13.4	N/A	N/A	N/A	N/A	N/A
1940	21.7	15.4	N/A	N/A	N/A	N/A	N/A
1950	20.6	17.6	N/A	N/A	1927	N/A	19
1960	17.4	19.5	N/A	249	3801	N/A	56
1970	13.8	25.9	24.3	330	6790	N/A	128
1980	13.4	32.0	35.1	345	9224	14066	350
1990	11.6	37.7	46.2	338	11308	17896	798
2000	10.1	40.0	49.3	352	12783	22308	1541

FIGURE 1.2

Selected indicators of American twentieth-century per capita consumption. Data from *Statistical Abstracts of the United States* and *Federal Reserve G19 Series*. Debt service ratios were never as scary.

refinement, retail store presentations ingeniously separate Americans from their money.

E-mail spamming is so cheap that a miniscule hit fraction pays off, but every American still receives about one piece of junk mail daily. Now, to gain a bigger piece of what's called "share of mind," advertising blurs more and more into entertainment with[66] the following:

- Long-form printed ads, which are hard to distinguish from other content in magazines
- Integration of product messages into entertainment programming
- Product placement in entertainment programming (often as props)

Wily persuasion is human nature, but for public relations professionals it became a commercial specialty. PR services run a gamut from damage control of corporate images to fogging public perception of scientific research findings.[67] Corporate ads finance the Internet. Online insurgents counterattack with fake messages. Pizzazz, not facts, grabs eyeballs. Is Adam Smith's invisible hand leading pigs to the trough?

However, physical consumption is consummated by mass credit condensed out of that mysterious ether called "the financial markets." Any future stream of income—home mortgages, credit card payments—can be collateralized and split into payment streams of various risk classes. Each stream can be marketed as a separate derivative package. Trading volume became colossal. Although only a fraction was directly tied to consumer spending, in 2005 global derivative trading billowed to $213 trillion of transactional froth, five times higher than the value of the global GNP; impressive numbers, but did this financial froth describe anything real and relevant?[68]

Derivative instruments traded are several degrees of abstraction removed from any activity whose value they purport to represent. Some have tangible property as collateral somewhere at their base; others do not. All derivatives derive value from someone, somewhere, promising to pay something, sometime—a system built on interlocking promises of streams of future income. They have different expected risks that underlying values will change, or that the payers will actually pay. Once specific contracts, like mortgages for example, are split into fractions, then bundled with similar payment promises into blocks for trading, untangling is difficult when nonpayments exceed the loss range projected by the models that set up the derivatives, as has been found in the subprime loan debacle. Of course, when some of the players originating the trading blocks in this system loaded them with risks beyond those assumed by the models, block trading values became uncertain, and trading locked up. Earlier, both Long-Term Capital Management and Enron illustrated the extremes to which deceptions—including self-deceptions—distort reality when this system runs amok. Finally in 2007, the whole system began to show signs that a global liquidity deluge was drying up into a liquidity drought. Whether financial confidence can be restored is uncertain at the time of this writing.

Concentrating on financial flops and monetary flows distracts from evidence that the physical edifice being financed has a shaky foundation. Overconsumption is the demand side of the resource limits discussed as Challenge #1. To consume just to be consuming—to create jobs and prop up industries because we don't know what else to do—is to adhere to an increasingly dysfunctional philosophy. Of course, a service economy—health insurance, child care, education—is now a big percentage of consumer transactions. However, service activities entail physical consumption, too—and thus continue physical expansion. As long as physical consumption can grow, this economic system's magic will promote it. Its core illusion is that more growth will overcome all problems, for example, that low taxes will free more investment to prime more growth, thus raising total tax revenue even higher, and motivating more people to solve the problems, and so on. That happened in the past, but it is also a rationale for deferring serious problems. Similarly, as is well known among business executives, this illusion fosters growth problems in fast-growing companies; that is, quantity-before-quality thinking.

Monetary transactions and valuations are indispensable, but we have to blow away the financial froth to see what we are physically doing. Doing this is often eye opening, as when personally looking at operations to create a flow chart of material through them when beginning a lean transformation. (The major long-run benefit is less in the new flow-path than in those who do the charting starting to see reality firsthand.) Likewise, summary indicators of aggregate physical phenomena, including their cumulative effects, present a different perspective even when the numbers are incomplete and approximate.

Were there no limits on resources, humans' capacity to consume could limit it (one can't eat twenty-four hours a day). The average adult who shops spends about two hours a day doing so.[69] Pervasive advertising conditions Americans to buy more, and to work more in order to shop more, but their shopping time is limited.[70] Online shopping may cut personal shopping time, but not consumption. Detached observers have noted that this culture of consumption is ridiculous in itself, but anticonsumption activists like Adbusters, with "Buy Nothing Day" and similar protests, have not dented it.[71] Others have noted that having more stuff saps more of one's time to keep it working if its quality and service are mediocre—and that from a customer's view, quality should prevail over quantity.[72]

Slower real growth may have showed up in conventional business analysis without anyone being conscious of Compression. For example, before the financial meltdown, Tom Osenton observed that organic growth in American corporate profitability in the last two decades of the twentieth century was slower than after World War II. Growth of corporate giants, like that of GE, became more characterized by acquisition and growth of financial services. Of course, it could just look this way in the United States because outsourcing of production to foreign countries was beginning to ramp up during this time. But most interesting is Osenton's tentative conclusion that at the saturation point of physical expansion, new innovations tend to cannibalize existing industries, not seed new growth.[73]

Cursory evidence that cumulative growth creates its own time-space limits is mixed. For, example, with millions more vehicles than licensed drivers and big urban traffic jams, flat auto sales in the United States should be no mystery. Airlines commoditized commercial flying long before September 11, saturating major airports. They keep flying more, but enjoying it less, with billions in losses. Speculation to prime dynamic growth in old utilities ended in the flim-flam that was Enron. Technological growth opportunities, like micro-tech comers in biotech or nanotech, aren't likely to lift an $14 trillion economy. More likely, they'll undercut fat old business models like big pharmaceutical companies.[74]

American national statistics show that, at best, per capita consumption has topped out, but has not begun a decline. An economic culture that measures success primarily by growth accumulates trouble for itself.

CHALLENGE # 4: PUSHBACK FROM THE HAVE-NOTS

In business and economics, globalization has come to mean the spread of free markets, corporate management, and financial systems throughout the world. To others, globalization may mean closer communication improving human understanding around the world. Business executives and workers who think that they have a winning hand in economic globalization generally favor it. Its critics protest social inequality, global corporate dictatorship, and environmental degradation—an extension of the fairness arguments surrounding capitalism since it began. Little unites them except detesting global corporatism, symbolized by the World Bank and International Monetary Fund (IMF).[75] Workers displaced by dog-eat-dog commodity competition, like unemployed textile workers everywhere, fume that globalization is crapping on them.[76] Moderate critics fear that global commercialism is washing their cultural heritage and moral values down a sewer. Harsh ones use phrases like "blood diamonds" to allege that privateers abuse workers who extract resources "from their own ground," inflame tribal feuding, and destroy local economies.[77] It's a strong version of small-town American protests that when Walmart moves in, it rips up the local economy with its social reciprocities. Pro-globalization pundits counter that economic growth will eventually benefit everyone, while decrying World Bank and IMF ineffectiveness promoting economic growth—expansion—claiming that policies favoring private investment can do it better.[78]

Antiglobal arguments resemble anticapitalist arguments dating back to chartered companies, the forerunners of corporations, used to institutionalize English and Dutch colonization. In a global corporate system, critics see the same benevolent arrogance that usually turned colonialism—by anyone—to cruelty in practice: population displacements that today we call ethnic cleansing; conscripted labor in various guises, including slavery or forced pauperization. They regard global corporate hegemony as one-sided trade rules disguised as democracy—*de facto* colonialism. Any hint of enforcing free-market doctrines by financial sanctions or military threats makes them livid.

A typical antiglobal rant is, "Militarism has become much more integrated into the workings of the global economy through international finance, trade, and investment regimes.... The new armies of occupation are the transnational corporations...providing security for investors, not citizens."[79] In industrial societies, antiglobal arguments protest things like a "corporate press" muffling dissent. In the developing world, plaints are more grassroots, like, "We've lost the basic freedom to drink water from the village well" (in other words, forced to pay for that which was free).[80]

Antiglobal pushback takes many forms that American media can seldom put into context. However, Hernando de Soto, a Peruvian globalization advocate, ominously captures the danger: "The hour of capitalism's greatest triumph is its hour of crisis."[81] Resentment and unrest easily lead to violence. While terrorism has many motives, it frequently boils out of groups that believe that they are not only culturally marginalized and disrespected, but also economically oppressed. Religious and ethnic differences sharpen the split. Although no one can fully explain the motivation of terrorists, almost everyone in the world was appalled by September 11, but not surprised by the targets. Wall Street and the Pentagon symbolize corporate world domination to far more people than al Qaeda.

De Soto's "danger" is that corporate expansion enthusiasts are blind to how their system is seen by others, building up huge reservoirs of resentment until the dams holding them break under the strain. Much of the world lacks the social and legal infrastructure of capitalism. Most occupied land does not certifiably belong to anyone. Property surveys or land titles, if they exist, are imprecise. Without clear land titles, courts frequently invalidate squatter claims of ownership. Without proof of ownership, the poor cannot collateralize any existing assets to generate liquid capital for a business. They lose out to the financially adept (and politically connected) who can leverage resources to get more. For example, this has happened on a large scale in China, creating unrest and rioting that pushed the Chinese government to enact new laws defining the rights and rules of private property.[82]

Have-not pushback is significant. Chinese development has displaced millions of people, uprooted socialized medical care, and barely begun to remediate environmental messes. So far, the backlash has been containable, but Chinese government policies had to address it.[83] In India, displacement for economic growth continues, although the Bharatiya Janata Party lost the 2005 election largely because the poor felt abused by it.[84] Complaints that economic growth is unshared, as in Panama, are common throughout the world.[85] In South Africa, even the African National Congress is accused of exacerbating income inequality.[86]

Inability to understand industrial operational requirements is also a problem. For example, uneducated Guatemalan small-holding farmers could not comprehend the schedules or quality standards demanded by supermarket buyers.[87] Poor prospects on the land drove people into cities, to make it however they could. Because of lack of opportunity and political instability, nearly 10 percent of the Guatemalan population has emigrated, and many more have tried. Many sneak into Mexico, where many Mexicans for similar reasons also head for a city or north to the United States. Coyote expediters thrive helping would-be emigrants flee to any country where money flows, with some hope of landing a poor-paying job. Thus the immigration issue in the United States is but one manifestation of a huge global rural-to-urban migration.[88]

A similar migration pattern is replicated all over the world. When they can no longer survive by their old ways on the land, the poor squat anywhere, notably urban slums, uprooted and disinherited from old customs, living on the ragged edge. They fall into street vending, cleaning for the rich, black-marketing—whatever it takes to stay alive. In environments like this, microfinance loans for cottage businesses let many poor people come closer to self-sufficiency, but alone it does not resolve issues like arbitrary eviction.

In older industrial societies, the mid-twentieth-century balm for this was manufacturing employment soaking up the labor displaced from farms. That no longer happens. Worldwide, manufacturing employment is not growing, but people are still being displaced from the land. Even China's official employment numbers show total manufacturing head count dropping until just after the turn of the twenty-first century; state-owned industries disgorged workers faster than private companies could add to payrolls. The notion of fast-growing, higher-tech sectors creating a big new manufacturing middle-class appears obsolete.[89] The United States' problem now is developing workers with broad and deep skills for a reduced manufacturing sector. (My personal experience visiting plants is that a high percentage of floor work is done by immigrants, not native-born Americans.)[90]

In the mid-twentieth century, Simon Kuznets proposed the theory that inequality rises when industrialization begins, but declines with continued growth, as more people obtain high-productivity work (or at least higher incomes).[91] A sign of a strong middle class is a relatively flat income distribution, sometimes measured with a Gini coefficient, as in Figure 1.3. Data from the development of old industrial economies supported this theory and bolstered capitalist faith in unlimited growth, eventually raising all boats. Figure 1.3 is a brief reexamination of Kuznets's theory.

Data support Kuznets's theory from industrial economies that grew more by developing people skills. Data from resource-rich countries do not. Brazil and Venezuela are high in income inequality. People in Sweden and Japan, at the low end, report much more equal incomes than in Brazil

| Country | Recent Gini Coefficient* | Gross Domestic Product Per Capita in Dollars | | | | Life Expectancy at Birth |
		In 1900	In 1950	In 1973	In 2003	
Japan	24.9 (109th)	1135	1873	11017	28000	82
Sweden	25 (108th)	2561	6738	13494	26800	80
United States	40.8 (41st)	4096	9573	16607	37800	77
Venezuela	49.2 (21st)	821	7424	10717	4800	74
Brazil	60.7 (3rd)	704	1673	3913	7600	71

* A higher Gini coefficient signifies higher income inequality. National ranking is in parantheses.

FIGURE 1.3
Five-country economic comparisons. Data from Nationstates and the *CIA World Factbook*.

and Venezuela. What's different about them? Both Brazil and Venezuela have been rich in products grown from the land or extracted from it. Both have colonial legacies and a history of landowner dominance, great ethnic diversity, and political instability. Sweden and Japan, however, are noted for modest natural resources, fairly homogenous ethnic populations—and relatively placid politics. That is, both Sweden and Japan had to depend on developing skilled people as their economic asset. Few people could get rich quick without bringing others along with them.

Even if some of us ever did live in Kuznets's kind of world, most of us did not and do not. The increasing imbalance between haves and have-nots is a global problem with severe consequences to everyone. Nigeria and Chad are prime examples. From 1970 to 1999, while oil revenues rose, Nigerian per capita income declined.[92] Chad has also discovered oil, and its social chaos is approaching that of nearby Sudan, another oil state whose Darfur region is in the sub-Saharan drought.[93]

Venezuela is particularly instructive. It struck oil early in the twentieth century, becoming the top producer in the world before Saudi Arabia began to pump up, and its oil was vital to allied efforts during World War II. By 1950, Venezuela's per capita income was fourth highest in the world. Oil production peaked in 1973. Now its per capita income is about fiftieth in the world. Oil revenue never translated into developing people to fully participate in a nonpetroleum economy. Caracas, the capital, is a city of about 5.5 million people, 80 percent of whom are in poverty. It got stuck as an ex-colonial, two-tier society.

Politically, Venezuelan liberals develop resources like oil slowly, keeping the price up to secure more revenue to benefit the poor (and maybe a few favored others). Conservatives, favored by all American governments, want to boost exploration and production to increase wealth now. Amid much rancor, Venezuela's president, Hugo Chavez, a liberal, was easily reelected in 2006. Prior liberal governments overcommitted social spending just as global petroleum prices declined. The oil industry blamed Chavez for derailing plans in 1998 to double Venezuela's oil output to seven million barrels a day by 2004, so Venezuela, like Saudi, could have opened the tap in case of shortfalls elsewhere. As seen by Amy Jaffe, chief energy analyst with Rice University's James Baker Institute, "Had they [oil companies] gotten there and stayed the course and everything had gone as planned, we would not be facing the kind of oil crisis we have today."[94]

Unsaid, but implied in that statement, is that the viability of the present global economic system rests on a bet that oil reserves are both sufficient and politically controllable enough that high prices will always reverse. It's been noted that the fortunes of "liberal, anti-global" governments rise with the price of oil and gas.[95] Instead, Chavez's oil diplomacy, for example, tweaks the United States' by selling cheap oil to energy-short countries and contracting with China to develop some of Venezuela's oncoming oil production, siphoning money into education and medical aid to Venezuela's poor.

Deniers have tried to fuzz numbers that suggest ongoing inequality. For example, in 2003 the *Economist* picked a couple of graphs from an inequality study to claim that the poor are catching up. They showed GDP per capita growing rapidly in India and China. However, GDP per capita averages away inequality *within* countries, and International Labor Organization data from *within* both China and India are mixed.[96] The percentage of people below $1 a day has dropped, but their total number is still many millions, and by 2007, *Economist* articles were beginning to acknowledge global economic inequality.[97]

In recent years, income inequality has increased in the United States, too. The high percentage of income by the top earners in the 1920s dropped during the 1930s depression; then after 1970 it began rising until in 2003 it was nearly as high as in 1928.[98] Recent data show income skewing even more sharply to the top, attracting political attention especially toward the upper 0.1 percent.[99] Many top earners are top sports or entertainment stars, so people are not overly upset with inequality if they think the top dogs merit it, and if they believe that some among them have a shot at the top rung—upward mobility. That belief is perhaps the core of keeping the American dream alive, so a politically bipartisan group decided to check it. Unfortunately, little evidence suggests that Americans' upward mobility is anything remarkable.[100]

Income status *relative* to the surrounding society, not the average for a society or GNP per capita, has been found to correlate with inequality in general health and to longevity, one marker of quality of life. This relationship is consistent with increasing evidence that overall health depends on social support and the psychology of social relationships. Economic haves' immune systems are stronger than those of have-nots. Economic haves are more likely to be integrated into society, confident, high in social trust, and usually nonviolent. Have-nots feel excluded, inferior, testy, and are possibly violent. If devoid of social approval, so that any contributions to society are unappreciated, they see little reason to take care of themselves physically, too, but live for the day. Such connections have been sensed for a long time, but have only recently been well documented.[101]

The internal support systems of have-nots' social groups degenerate, especially if alien systems they don't understand are imposed on them. Have-nots may even feel out-of-place by aspiring to perform by the expectations of economic haves, one reason why upward mobility is problematic.[102] Industrial society haves easily interpret these attitudes as inherent indolence or incapacity, assuming that anyone seeking admission to *their* social milieu must demonstrate responsibility using their cultural rules.

The magnitude of this issue is better appreciated by reviewing the expansion of the world's cities. In 2007, about 3.3 billion people, over half the global population, lived in cities. About 0.7 billion were in one of the largest 100 cities. At the present rate of global migration into cities, within a few years the global rural population will start to shrink. About a billion people, a third of the global urban population, live in what the United Nations defines as a slum according to one of five conditions: inadequate protection from weather, no close access to safe water, more than three people per room, no access to a sanitary toilet, and insecurity of tenure. Insecurity of tenure means that no legal sanction forbids capricious eviction—by *anyone* else who covets the space, anytime they may want it. By 2020, the global urban population is projected to grow to about 5 billion.[103]

For example, the population of Mumbai, India, is about the same as New York City—20 million or so—but its population density is about eight times greater, and 5 million of its people live in slum conditions. Given

earth's resource constraints, it is obvious that adequate physical housing for the world's slum dwellers will take some out-of-the-box thinking.
Global physical development with human development is a super-sized
problem, for it seems obvious that any real solution has to come from
enabling the slum dwellers to pull themselves up. If we are to provide a
reasonable quality of life to a global population in Compression, then the
mess of the urban poor is one of the major messes to untangle, starting
with unraveling why so many of them want to migrate to a city.

Mere numbers cannot describe the human conflict associated with this
backlash. Amy Chua has compiled many examples in which economic
inequality stirred up with ethnic and religious animosity brewed violence. In
some cases, minority business elites were in cahoots with autocratic governments, as with Siaka Stevens and the Lebanese in Sierra Leone; or Ferdinand
Marcos and the Chinese in the Philippines (where the phrase "crony capitalism" originated). In other cases, a populist autocrat displaced a minority
elite, like Robert Mugabe in Zimbabwe. Or in the Venezuelan case, an elected
leader like Hugo Chavez riled the old economic elite.[104] Unfortunately, if
resentment of elitist approaches to development rejects professionalism,
as has happened in Fiji, basics such as health care and education suffer.[105]
Improving quality of life involves more than material goods. Different kinds
of working organizations have to evolve to take on such challenges. Most of
today's enterprises, public or private, do not seem up to them.

Unfortunately, unless people see a bigger mission in life than disputing
their inequalities or human differences, they chew up such quality-of-life
challenges in emotional meat grinders, crushing all hope of viewing them
objectively or dealing with them imaginatively. The current system of business seems particularly prone to this. Beholden to ownership interests, it
lacks imagination and empathy with many of the people to be developed
to improve their quality of life. Their supposed beneficiaries are likely to
see their tactics as using democracy as a facade to allow wealthy dominant minorities, fearing chaos, to manipulate elections. Unwittingly, the
International Monetary Fund (IMF), World Bank, and other institutions
finance environmentally destructive projects, abet corruption of local cultures, and fail to let locals learn to fish on their own. Noreena Hertz explains
how the global financial system keeps poor countries perpetually indebted
somewhat like slave indebtedness.[106] John Perkins, economic insider turned
whistleblower, describes pressuring locals by manipulating elections, playing financial hardball, and as a last resort, using military force.[107]

Pushback by have-nots continues endless centuries of fairness conflicts. Whether we can overcome old competitive instincts to resolve such issues is dubious, but we must try. Right now, China, with its seemingly endless supply of cheap labor and boundless capital-spurring growth, has also spurred a great deal of pushback from the Chinese displaced by it. When cost competition with Chinese-made commodities seems hopeless in the industrial economies, those displaced there want currency fairness, tariffs, and so on—the usual remedies.[108] But considering global population trends, China and India hold an ocean of have-nots at least four times bigger than all the older industrial economies combined, while these old societies have their own populations of have-nots.[109] We have entered a postcolonial global transition. Prospects of minimizing pushback depend on finding ways to develop have-nots to improve their own quality of life without destroying that of others.

Can we unify around a bigger common human mission than those that divide us? If so, there's a chance, however slim. Otherwise, resentful have-nots may destroy a lot more ecology on their way out. Thus far, billions of people have seen more garbage than benefits trickling down from consumer societies, including the newest ones like China. Globalization has to be redefined as growth in quality of life rather than growth in quantity of consumption—unending expansion of transactional commerce.

Unifying everyone on earth seems hopeless, and no self-appointed genius is apt to come up with a miracle solution. It's much wiser to develop working organizations that can take leadership multiplying the human capacity to actually cope with Compression.

CHALLENGE #5: SELF-LEARNING WORK ORGANIZATIONS

All descriptions of the first four challenges were out of date before they were done. Each is a near-infinite mix of ever-changing, interrelated conditions. Just comprehending the enormity of these challenges taxes the mental capacity of everyone who grew up in an affluent society in which expansion was considered normal—without ever thinking much about it. But once past this point, the question is what we are going to do about it.

We have to start doing very differently, very quickly, and on a wide front. A few tweaks of the current business system by government policy are insufficient. Government policy has long supported expansion by opening various areas to development by a business system whose decision logic is generally based on expansion. That system is rapidly becoming the victim of its own extravagant success.

Doing much more using much less is a different kind of mentality. The human discipline to do something this different is not trivial. Will we continue to civilize ourselves and step up to these challenges, or regress to fighting for survival separately?

Squabbles over fishing limits for the bluefin tuna illustrate the difficulty of this shift in thinking. The near-collapse of bluefin tuna fishing follows a script similar to that from many collapses before it. Marine biologists and fishing commissions lack precise data, but what they do have points to impending collapse. But to fishermen with boat loans to pay and families to feed, discontinuing fishing is complete economic collapse. They push to stretch a fishing limit as far upside as possible. Some of them poach illegally, causing reported catch numbers to be suspect, and they discount the advice of so-called experts who do not actually fish. Much like fishermen involved in cod, striped bass, and other fishing collapses before, fishermen will compete against each other to economically survive until there is nothing left for any of them.[110] The tuna fishers have what's known as a wicked problem because they do not have a constitution to which they are committed, and that provides for finding facts, weighing evidence, and overcoming fear to reach a decision that all can accept, including alternatives for those whose means of livelihood are damaged by the decision.

It's not hard to see the parallels between the 2008 global financial debacle and a fishery collapse. Financial institutions found—once again—that they had competed to blow another bubble, but when it popped, gum covered everybody's face. Rethinking what they are doing does not go much beyond prescriptions to reprime growth with tighter control of the bubbles. Few see that most of their expansionist assumptions are larded with no end of self-justifying rationales to continue prodding physical expansion of the global economy by the current system. One has to stand outside this system to recognize that it is destroying itself. Expansion may be reprimed for another round or two, but it is trying to do something that it cannot do much longer. However, by adopting very different basic thinking, it could preserve itself in some form.

Expansionism is making an ever bigger pie, which some can hog while leaving enough for all. Even if allocation is contentious, when there are plenty of goodies to share, managing both companies and politics is easier. With resources to spare, we can recover from many mistakes. When all resources must be carefully conserved, we have to think carefully, making fewer mistakes—and smaller ones. Creating a smaller-resource pie of higher quality, aiming to let every responsible person have a livable slice, is a revolution (recognizing that in every society not everyone is socially responsible). This revolution is so fundamental that it is not a conventional street-protest, political revolution. It redefines what work is and what a work organization is supposed to do, so it is more like a rapid evolution by working organizations and those who work in them.

MEETING THE CHALLENGES

Thus Compression refers to two phenomena. The first is the physical economy's squeeze on the earth's resources and ecological viability. The second is how to deal with the first squeeze by developing people and organizations to squeeze (or compress) their work processes, products, and use cycles of their customers, which is the main subject of this book. Compression in this sense is necessary just to deal with increasing complexity if we had no resource shortages or environmental constraints.

The challenge of developing people everywhere, in all cultures, to cope with Compression is overwhelming. One of the few feasible ways to do it is by morphing our key work organizations into a form that can transform how much of the world's work is done. To do that, they have to fulfill two missions:

Make radical, major improvements in performing work that today consumes most of the world's energy and materials, generates most of its toxic substances, and primarily influences the quality of human life.

Coach the rest of us on how to better cope with Compression; that is, take the lead in enabling the world's population to also deal with the challenges of Compression.

During expansion, almost everyone on earth became a customer of some of these organizations. At present, these work organizations may be classified as for-profit, nonprofit, or government; but in Compression, these distinctions blur. The for-profit ones have been very persuasive in stimulating people to expand consumption. Perhaps by hands-on instruction as well as persuasion, they can become even more ingenious in helping us all learn how to live well while using a fraction of the resources we consume today.

So this book is addressed to the leaders of today's work organizations: profit, nonprofit, government, or some hybrid. Some are considered fairly advanced in proficiency, but even for them Compression requires a revolution in leadership to develop the work capabilities and responsibilities of participants, blending them to do what no one person can do alone. These organizations must learn the right things to do at the right times, and how to execute them almost perfectly.

Vigorous Learning Organizations

This new work organization is labeled a vigorous learning organization. When extended to cover a network of suppliers, customers, and other partners, it can be called a vigorous learning enterprise. *Vigorous* is intended to imply energetically doing things, like transportation, health care, remanufacturing, and so on, so that *learning* is not confined to academic exercise. We must change what we do and how we do it, not merely what we study or what we invest in. A rough estimate is that such organizations might cover as much as 40 percent of the present United States workforce—a change broad in scope, but not unthinkable. This book is addressed to those who may take the challenge of creating such an organization.

The ideas for vigorous learning organizations amalgamate the best seen in actual practice today, and then nudging them further, for a vigorous learning organization is a big leap from our best today. It needs a mission that copes with Compression. To carry it out, it must be oriented toward processes or systems rather than finance. To make major changes faster than ever before, and with few mistakes, people working in it need a common learning methodology and a learning system. They must be more than empowered; they must become professional in social dedication as well as in technical proficiency.

Emotional and behavioral development is probably the hardest aspect of creating a vigorous learning organization. Today, lean operations or Six Sigma are only beginning steps toward this vision, but even so, experienced leaders realize that human problems are the biggest obstacles to implementation. From alpha instincts to petty jealousies, emotions impede cooperation for mutual learning. The behavior to counteract this is noninstinctive. It has to be learned and reinforced regularly. Instilling discipline for vigorous learning is no light matter, but it is very different from motivating people with carrot-and-stick incentive systems.

Health Care as an Example

When consumed by marketing and financial competition—growth—health care organizations can't picture how things actually work from the green-gown level up to the total system level. Each party can pretend that if it is making money, the total system must be doing the best it can. But patients and most workers in health care see that the system is a bureaucratic mess. Its strong point is new biotech miracles, but these suck up resources and drive up costs—a mixed blessing.[111]

New products, tests, or pharmaceuticals added to the system without displacing much else gradually complicated it, bogging it down in cost-hassling, episodic medical interventions in which communication of massive case detail is cumbersome. To escape the insurance morass, many physicians flee to fields like cosmetic medicine for people with bucks. Much of the waste is self-created imaginary complexity, add-on after add-on to the system without simplifying anything until it becomes ineffective at meeting society's direst needs.[112] Health organizations are working to decrease medical errors and to improve the workflow of patients, lab tests, and paperwork, but that does not reduce business incentives to grow more revenue from health care. Only reworking the system to support quality over quantity at the green-gown level can touch that.

Rethinking Expansionary Assumptions

This seems strange because expansionary thinking has prevailed for so long that its hidden assumptions are deeply embedded in conventional business thought and economics. Compression forces us to go back to where everything is simple again, and rethink why we do it.

For example, most businesses emphasize return on investment. Why? Because investors seek maximum return on their money. And why? One reason is that it is a legacy from nineteenth-century railroads that raised enormous sums of money in nearly unregulated capital markets and had to give investors a good return to avoid being raided. But if we are investing in a health care operation, why seek a maximum return on investment? That is not really the mission of a health care operation.

In Compression, human capability and organizational performance to mission is the main objective. Capital is only the monetary valuation of many of the things with which people work. In Compression, leadership of people, work processes, and learning processes are what really count in improving performance to mission. As can be seen in the recent financial system meltdown, organizations such as banks, auto companies, airlines, and health services really have positions of public trust. If they go bust, not only must everybody help recapitalize them, but more importantly, all of us must rely on their performance. The rationale that top managers' first priority is to maximize return to owners has to go. It subverts leadership of a mission-centered organization to benefit all its stakeholders.

So there is a lot to rethink. Criticism of expansionary business conventions may seem heavy-handed, but to cope with Compression, leaders need to expand their thinking. Pioneering books that broached similar concepts used terms such as *Natural Capitalism* (the resources and functions that nature provides to us) that business thinkers can relate to; but to get on with it, leaders need a broader view of their responsibility.[113] Business leaders need to learn much faster, which is fearful at first. To them, financial failure is a much more imminent threat than global physical collapse. The turning point comes when they realize that dealing with Compression is not a nice-to-do image booster when we have enough money. There will never be enough money. Coping with Compression is a different economic philosophy with different rules. We have to use whatever we have to do what we must.

Beginning about 1980, to meet Japanese competition, American companies took up operational excellence with programs like Six Sigma and blizzards of acronyms like TQM, JIT, and QFD. This did have an effect. For example, vehicles of today's quality could be neither designed nor built with the practices of 1980, even if the technology had advanced. Computers, software, and technology helped, but as in music, virtuoso techniques do not add up to great performance unless masterfully blended.

This excellence revolution still struggles against the assumptions of financial guidance. Everything is seen as a trade-off, not as a self-reinforcing whole: Better quality must cost more. Fast delivery and prompt service must cost more. Actually talking with customers must cost more. Cleaning up pollution must cost more. But companies that violate these nostrums get awards, and as a side benefit, usually make more money, too.[114] Fear of losing out by the present system is a big ogre hindering its transformation.

TAKING ON THE CHALLENGES OF COMPRESSION

Physical laws and nature's resources do not negotiate with any market consisting of human valuations. We only negotiate among ourselves, but we affect nature and its resources by what we do. Each of the challenges of Compression has been more eloquently presented by others. They are all critical, and some conclude that industrial society may be destroyed by Challenge #4, pushback, sooner than by any of the others. At stake is no less than whether we meet these challenges by rising to a new level of civilized achievement, or whether many achievements of the past collapse in chaos like those of most ancient civilizations.

We need more than patchwork on the expansionary system. We need vigorous learning enterprises with a guidance system fundamentally different from business as usual. Some decisions—many more than today—are too important to make by financial calculus. We need a process theory of value to offset monetary values, derived from market exchange. A process theory of value has to rate or prioritize the importance to us of both manmade and natural processes where market values are impossible or irrelevant to assess. We have to consider the effects of what is done, not what we may possess.

The second challenge is overcoming something here dubbed "tribalism." Humans appear hard-wired to separate into groups to squabble over all kinds of differences. If we are going to agree on the facts and logic about anything important, we have to overcome this instinct, in essence agreeing on how to agree. In government, this is called a constitution. Vigorous learning organizations need to progress toward a global learning constitution—a working framework of rules and customs by which all of us can improve quality of life as much as possible using limited resources.

For clarity of challenge, arbitrarily quantifying the definition of Compression emphasizes that it cannot be reached without a major overhaul of the economic rules of engagement: Assure for everyone in the world at least the quality of life as that now in industrial economies while using less than half the mass of virgin raw materials, less than half the fossil fuel energy, and having zero toxic releases—quality over quantity always. None of us can fix this in mind without dislodging prior concepts about how the world ought to work.

ENDNOTES

Extended version of the footnotes available at http://www.productivitypress.com/compression/footnotes.pdf.

1. Thomas Malthus' original paper is "An Essay on the Principle of Population," 1798.
2. According to the U.S. Geological Survey: http://minerals.er.usgs.gov/minerals.
3. M. King Hubbert, "Energy from Fossil Fuels," *Science*, Feb. 4, 1949. Also Kenneth S. Deffeyes, *Hubbert's Peak; The Impending World Oil Shortage*, Princeton University Press, Princeton, NJ, 2002.
4. Around 6 percent, including the energy to process the feedstock (per the German Chemical Industry Association).
5. Deffeyes, Odum, and other authors estimated the yield of the first gushers in this range.
6. Hubbert, Deffeyes, and other analysts note small and low energy-yield discoveries only push out the year for peak oil; they do not reverse the trend.
7. Colin J. Campbell and Jean Laherrere, "The End of Cheap Oil," *Scientific American*, March 1998, pp. 78–83. Well-reasoned articles (and less-reasoned ones) can be found at http://www.peakoil.net.
8. Jean Laherrere, "Future of Natural Gas Supply," *ASPO 3rd International Workshop*, Berlin, May 24–25, 2004.
9. Jad Mouawad, "Violence in Nigeria Sends Oil Prices Higher," *New York Times*, Feb. 21, 2006.
10. "Biofuels in the U.S. Transportation Sector," Energy Information Administration, DOE, Feb. 2007, at www.eia.doe.gov/oiaf/analysispaper/biomass.html.
11. Robert F. Service, "Framework Materials Grab CO_2 and Researchers' Attention," *Science*, Feb. 15, 2008. Technologies capable of absorbing CO_2 are being developed, but scale-up is a big problem.
12. Commercial PV cells are now 15 to 20 percent efficient at converting solar energy to electricity; their energy yield is still below 1. Higher conversion efficiencies and simpler production methods are in development.
13. An example article on this is "Manila Rice Pleas a Wake Up Call for a Hungry World," Reuters (in the *Daily Times*, Pakistan), Feb. 28, 2008.
14. Leo Horrigan, Lawrence S. Wallace, and Polly Walker, "How Sustainable Agriculture Can Address the Environmental and Human Health Harms of Industrial Agriculture," *Environmental Health Perspectives*, May 2002. The average U.S. farm uses about 3 calories of energy to grow 1 calorie of food energy. If energy is used to grow fuel, its overall energy yield is cut substantially.
15. Food and Agriculture Organization of the United Nations (FAO): http://faostat.fao.org.

16. Joseph J. Romm, *The Hype about Hydrogen*, Island Press, Washington, D.C., 2004. A truly novel breakthrough is necessary to commercialize hydrogen on a large scale and with a high energy yield.

17. Jared Diamond, *Collapse*, Viking, New York, 2005, and similar later sources.

18. Joseph A. Tainter, *The Collapse of Complex Societies*, Cambridge University Press, Cambridge, U.K., 1988.

19. Peter Turchin, *War and Peace and War*, Plume (Penguin Books), New York, 2007.

20. From Marq DeVilliers, *Water: The Fate of Our Most Precious Resource*, Mariner Books (Houghton-Mifflin), New York, 2000.

21. Percentages from U.S. Geological Survey: http://ga.water.usgs.gov/edu/earthwhere-water.html, and originating from Gleick, *Water Resources*, 1996.

22. According to the British Pump Manufacturers Pump Industry Association Sustainable Energy Strategy – Executive Summary, August 2007.

23. The gist of the Cochabamba story is in the archives at *Source Water and Sanitation News*, http://www.irc.nl/page/2082.

24. Typical measures to regulate over-pumping can be found at Texas Alliance of Groundwater Districts: http://www.texasgroundwater.org.

25. A Western dictionary definition of *usufruct* is "the right to use the profits and advantages of something belonging to another as long as the property is not damaged or altered in any way."

26. It took about 1000 years to stop Christians from rechristening "holy" wells after saints.

27. Garrett Hardin, "The Tragedy of the Commons," *Science*, 162, 1968, pp. 1243–1248.

28. Lester R. Brown, *Outgrowing the Earth: The Food Security Challenge in an Age of Falling Water Tables and Rising Temperatures*, W.W. Norton & Company, New York, 2005, pp. 106–107.

29. According to the Edinburgh, Scotland, News On Line, Apr. 5, 2005.

30. Jane Perlez, "In Life on the Mekong, China's Dams Dominate," *New York Times*, Mar. 19, 2005.

31. Google "Aral Sea" and "Novaya Zemlya" to see hundreds of hits varying in credibility.

32. United States Geological Survey, "The Future of Planet Earth: Scientific Challenges in the Coming Century," Feb. 2000. Summary at http://www.usgs.gov/newsroom/article.asp?ID=653 (a regularly update site).

33. The Singapore desalinization plant supplies about 10 percent of its water and takes 4.2 kWh per cubic meter of water. Information from Black & Veach at http://www.bv.com/wcm/press_release/04072006_8521.aspx.

34. Ian Sample, *Guardian Unlimited* (UK), "Scientists Offered Cash to Dispute Climate Study," Feb. 7, 2007.

35. B.H. Keating and W.J. McGuire, "Island Edifice Failures and Associated Tsunami Hazards," *Pure and Applied Geophysics*, 2000, pp. 899–956. A follow-up was Steven N. Ward and Simon Day, "Cumbre Vieja Volcano—Potential Collapse and Tsunami at La Palma, Canary Islands," *Geophysical Research Letters*, 2001, pp. 3397–3400.

36. The Tsunami Society (http://tsunamisociety.org/) references the "debunking" paper at www.drgeorgepc.com/TsunamiMegaEvaluation.html.

37. The *2007 Benfield Hazard and Risk Science Review*, p. 23.

38. Jeffrey Lacasse and Jonathan Leo, "The Media and the Chemical Imbalance Theory of Depression," *Society*, 2008, pp. 35–45.

39. Constance Holden, "Report Warns of Looming Pollination Crisis in North America," *Science*, Oct. 20, 2006.

40. The Mid-Atlantic Apiary Research and Extension Consortium at Penn State is the designated coordination center for the CCD Working Group.

41. "Plight of the Bumblebee," *New Scientist*, Upfront Section, 15–21 July, 2006 (news report of a study appearing in *Journal of Applied Ecology*, DOI: 10.1111/j.365-2664.2006.01199.x).

42. A summary of frog disappearance is at http://www.wettropics.gov.au/pa/pa_frog_crisis.html.

43. Elizabeth Rosenthal, "Near Arctic, Seed Vault is a Fort Knox of Food," *New York Times*, Feb. 29, 2008.

44. Enter a search on http://www.antarctica.ac.uk "The Ozone Hole" for reports, pictures, graphs, etc.

45. Census of Marine Zooplankton, www.cmarz.org. Quick summary at http://news.mongabay.com/2006/0504_coml.html.

46. John Roach, "Source of Half of Earth's Oxygen Gets Little Credit," *National Geographic News*, June 7, 2004.

47. Stephen Jay Gould and Niles Eldridge, "Punctuated Equilibria: An Alternative to Phyletic Gradualism," *Models in Paleobiology*, Freeman, Cooper and Company, San Francisco, chap. 5, pp. 82–115. Can be found at: http://www.stephenjaygould.org/library.html.

48. The 6 percent is from NASA Goddard, Sept. 24, 2003, at http://www.nasa.gov/centers/goddard/news/topstory/2003/0815oceancarbon.html. Also Victor Smetacek and James Cloern, "On Phytoplankton Trends," *Science*, March 7, 2008.

49. David Appell, *Scientific American* (In Focus Section), Aug. 2, 2004.

50. Reports of low plankton blooms continue, but understanding is too limited to create a stir in the media.

51. Reefs might or might not adapt to a pH change, according Prof. Stephen Palumbi, Stanford University. See: http://www.eurekalert.org/pub_releases/2009-05/su-ssf051909.php

52. The term "wicked problem" originated from Horst W.J. Rittel and Melvin M. Webber, "Dilemmas in a General Theory of Planning," *Policy Sciences* 4, 1973, pp. 155–169. See also Strategy Kinetics at http://www.strategykinetics.com.

53. From the World Health Organization 2007 Report: http://www.who.int/whosis/whostat2007/en/index.html.

54. From a graph by the Center for Sustainable Systems, University of Michigan: http://css.snre.umich.edu/css_doc/CSS04-15.pdf.

55. Two models are at http://dieoff.org/page110.htm and http://www.footprintnetwork.org/en/index.php/GFN/page/methodology.

56. James Lovelock, *The Revenge of Gaia*, Basic Books, New York, 2006, p. 141.

57. Estimates based on data from both *Ward's Automotive* and the Motor Vehicle Manufacturers Association.

58. Done for fun based on figures from *Ward's Automotive* and looking at a Wal-Mart parking lot.

59. Estimated from Table HS-40, *Statistical Abstract of the United States, 2003*, Mini-Historical Statistics.

60. *Statistical Abstracts of the United States 2003*, Table 994, and Table HS-40, from its Mini-Historical Statistics.

61. Denny Lee, "Taking the Garage Along for the Ride," *New York Times*, Jan. 14, 2005.

62. Estimates calculated from various tables in *The Statistical Abstract of the United States, 2003*.

63. Niel deGrasse Tyson, "Energy to Burn," *Natural History*, October 2005.

64. *Statistical Abstracts of the United States*, 2009, Table 1132 and the series from the Federal Reserve (www.federalreserve.gov/releases/housedebt/default.htm) do not give the same impression due to the way data is presented.
65. From McCann-Erikson, the *Statistical Abstract of the United States, 2003*, and World Bank income per capita.
66. To see how pervasive this is becoming, check http://www.madisonandvine.com.
67. Anthony Pratkanis and Elliot Aronson, *Age of Propaganda*, Henry Holt, New York, second edition, 2001.
68. International Swaps and Derivatives Association: http://www.isda.org/index.html (ISDA Annual Market Survey).
69. U.S. Department of Labor survey, Table 1: http://www.bls.gov/news.release/atus.t01.htm.
70. Juliet Schor, *Born to Buy: The Commercialized Child and the New Consumer Culture*, Scribner, New York, 2005.
71. See www.adbusters.org.
72. James P. Womack and Daniel Jones, *Lean Solutions: How Companies and Customers Can Create Value and Wealth Together*, Free Press, New York, 2005.
73. Tom Osenton, *The Death of Demand*, Financial Times Prentice Hall, Upper Saddle River, NJ, 2004.
74. See for example, "Category Killers: 5 Nanotechnologies That Could Change the World," *Forbes/Wolfe Nanotech Report*, from Angstrom Publishing, a subsidiary of Lux Capital, New York, September 2004, p. 3.
75. Wikipedia's overviews of Anti-Globalization and related movements is up to date, and as expected, disputed. (See http://en.wikipedia.org/wiki/Anti-globalization)
76. James Brooke, "Down and Out in Mongolia," *New York Times*, Dec. 29, 2004.
77. Larry Richter, "Diamonds' Glitter Fades for a Brazilian Tribe," *New York Times*, Dec. 29, 2006.
78. For example, see Joseph E. Stiglitz, *Globalization and Its Discontents*, W.W. Norton, New York, 2002.
79. Passage from "Re-Criminalization of Dissent," notes from Concordia University Conference, March 2002.
80. Saritha Rai, "Anti-Globalization Forum Adds Variety of Causes to Its Agenda, *New York Times*, Jan. 20, 2004.
81. Hernando de Soto, *The Mystery of Capital*, Basic Books, New York, 2003, p. 1.
82. Li Gao, Jonathan Lindsay, and Paul Monro-Faure, "China: Integrated Land Policy Reform in a Context of Rapid Urbanization," World Bank Report, Agricultural and Rural Development Notes, February 2008.
83. Joseph Kahn, "China's Haves Stir the Have Nots to Violence, *New York Times*, Dec. 31, 2004, and "China Worries about Economic Surge That Skips the Poor," *New York Times*, Mar. 4, 2005.
84. For example, Prafulla Das, "Tribals Vow to Oppose Displacement," *The Hindu* (online newspaper of India), Jan. 3, 2007.
85. "Panama: Poverty Despite Economic Growth," *Prensa Latina*, Feb. 1, 2007.
86. John Pilger, *Freedom Next Time*, book manuscript reviewed by *Al Jazeera*, Aug. 29, 2006.
87. Celia W. Dugger, "Supermarket Giants Crush Central American Farmers," *New York Times*, Dec. 28, 2004.
88. Vanessa Burgos, "Mexico-Guatemala—the Other Border," originally in *Upside-Down World*, Nov. 14, 2007. Global rural-to-urban population shift data come from

the U.N. Department of Economic and Social Affairs Population Division "World Urbanization Prospects" series.
89. Comparisons from a U.S. Bureau of Labor Statistics Report at Mfg employment Comparative Civilian Labor Force Statistics, http://www.bls.gov/fls/flslforc.pdf and similar sources.
90. *Manufacturing & Technology News*, "Manufacturing Educator Pleads with Industry to Help Recruit Students," Feb. 28, 2008.
91. Simon Kuznets, "Economic Growth and Economic Inequality," *American Economic Review*, 45, 1955, pp. 1–28.
92. Michael L. Ross, "Nigeria's Oil Sector and the Poor," Report to the U.K. Government, May 23, 2003.
93. See for instance, Carmen Gentile, "Analysis: Chad Unrest May Be over Oil," *UPI*, Feb. 7, 2008.
94. Greg Flakus, "Analysts Concerned Venezuelan Turmoil Affecting Oil Prices," Voice of America News, www.voanews.com, as ported to EIN News (http://www.einnews.com), Sept. 12, 2004.
95. One such observation is Thomas L. Freidman, "The Oil-Addicted Ayatollahs," *The New York Times*, Feb. 3, 2006.
96. The *Economist* article was "Catching Up," August 21, 2003. James Galbraith composed the most popular exposé of the *Economist's* gaffe, http://salon.com, March 2004.
97. For example, "In the Shadow of Prosperity," *The Economist*, Jan. 20, 2007.
98. Thomas Piketty and Emmanuel Saez's studies are widely accepted because they are methodologically thorough. See "Income Equality in the United States, 1913–2002," at http://elsa.berkeley.edu/~saez/piketty-saezOUP04US.pdf.
99. Data are in 2007 papers cited on Saez's Web site: http://elsa.berkeley.edu/~saez.
100. Elizabeth Sawhill and John E. Morton, "Economic Mobility: Is the American Dream Alive and Well?", can be found at http://www.economicmobility.org/assets/pdfs/EMP%20American%20Dream%20Report.pdf. Much earlier studies conclude the same, for example, Robert Erickson and John H. Goldthorpe, "Are American Rates of Social Mobility Exceptionally High?" *European Sociological Review*, 1985, pp. 1–22.
101. Two books explaining this are Richard Wilkinson, *The Impact of Inequality*, The New Press, New York, 2005; and Michael Marmot, *The Status Syndrome*, Times Books (Henry Holt), New York, 2004.
102. Donna Y. Ford, Tarak C. Grantham, and Gilman W. Whiting, "Another Look at the Achievement Gap," *Urban Education*, Feb. 2008.
103. U.N.-Habitat *State of the World's Cities, 2006/7*, at http://ww2.unhabitat.org/media-centre/sowckit2006_7.asp.
104. Amy Chua, *World on Fire*, Doubleday, New York, 2003.
105. Shailendra Singh, "Island Populations Thinning Out from Migration," Inter Press Service News Agency (www.ipsnews.net) Jan. 2, 2007.
106. Noreena Hertz, *The Debt Threat*, HarperCollins, New York, 2004.
107. John Perkins, *Confessions of an Economic Hit Man*, Berrett-Koehler, San Francisco, 2004.
108. For example, *Manufacturing & Technology News*, Feb. 28, 2008; "Hunter and Ryan Hope Press Corps Can Help Convince Congress to Take on China."
109. Poverty levels range from 6 percent to 80 percent, using poverty levels in country profiles data from the *2007 CIA World Factbook*: https://www.cia.gov/library/publications/the-world-factbook/fields/2046.html.

110. World Wildlife Federation News Center, "Tuna Commission Comes up with a Disgrace, Not a Decision," Nov. 24, 2008; in the archives at www.panda.org.wwf_news/
111. Shannon Brownlee, *Overtreated*, Bloomsbury, New York, 2007.
112. Nam P. Suh, *Complexity*, Oxford University Press, New York, 2005.
113. Amory Lovins, Hunter Lovins, and Paul Hawkens, *Natural Capitalism*, Back Bay Books (Little, Brown), New York, 2000.
114. Baldrige winners and conspicuous "lean performers" have consistently outperformed the S&P 500 index. So has the Global Lamp Index (60 companies taking a "biological living system" approach to the triple bottom line).

2

Learning from Toyota[1]

Toyota is not a vigorous learning enterprise coping with the challenges of Compression—at least, not yet. It's not very different from Honda or several of its suppliers, but it must be the world's most studied company. What attracts attention is its consistent quality of vehicles, Toyota Production System (TPS), Toyota New Product Development, pioneering of hybrid drive vehicles, and its huge pile of cash nicknamed "Bank Toyota." It became a victor in the global economy in expansion.

For three decades, Toyota performance has stood out so that many other companies would like to discover its secrets. They can't because Toyota hides most of them in plain sight, and few executives have the patience to deeply understand what they see.

Most companies develop assets. Toyota develops people—all people, not just a few fast-track stars. Every employee is expected to become a professional, seeing and solving problems and collaborating with others doing it. Because of consistency doing this, Toyota does many other things very well. It does not trot out management fads or use gimmicky incentive plans.

Toyota strives for a never-attainable vision of perfection, zero unsatisfied customers, a goal that in the United States Toyota calls "True North." It regards all employees as professionals capable of heading True North in their daily work. It strives to develop the full potential of each employee to see and solve any kind of problem that diverts work energy from True North. Both individual and group problem-solving processes are structured so that thinking is clear and disciplined. Best practices and best solutions are documented in a knowledge system now partly computerized so that attention concentrates on solving new problems, not re-solving old ones. And it respects the environment. A green leaf is one of its logos (although selling a big Tundra pickup truck seems inconsistent with it).

THE TOYOTA WAY

Most manufacturing executives around the world know of TPS, but mostly as a set of industrial engineering techniques to eliminate waste and smooth the flow of operations. Fewer realize that TPS is designed to prompt all employees to see process problems readily and solve them promptly.

All automotive executives also know that Toyota develops new vehicles quickly, and with very few design flaws. Just as TPS trickled out earlier, the main features of Toyota's New Product Development system (TNPD) are emerging. But few executives realize that TNPD is intertwined with a system of organizational learning that is among the best in the world, so well done that many research organizations could get some pointers from it. The bedrock for all of this is a few principles called the Toyota Way.

The principles of the Toyota Way are frugally encoded on one printed page, which the company will not officially release. Toyota prefers that their people absorb the principles thoughtfully and carefully, by living them and by being mentored in them. If posted on the walls as general reading, the Toyota Way would decay into corporate eyewash. Learning to live the Toyota Way is thereby a very important, but tacit, principle of it.[2] However, Toyota does publicly proclaim the two major pillars of the Toyota Way: (1) respect for people (individually and collectively) and (2) continuous improvement.

The second pillar is well known to anyone studying TPS. The pillar too often unseen is respect for people, equally important. TPS techniques are easily seen. They can be described in words and diagrams. But Toyota's intent in using TPS techniques is to create within people a different pattern of thinking at work—rigorous and disciplined, but actionable. Today, Toyota calls TPS its Thinking Production System.

The Toyota Way applies to all activity, not just production; and as Toyota notes, no words can fully encompass it. Despite this caveat, a tentative Western view of it might have five major elements:[3]

- Respect other people; share your knowledge with them; build trust.
- Stimulate professional development of yourself and others; share opportunities; maximize others' performance before your own.
- Create organizational learning using facts and evidence; whenever possible, go observe the facts of a situation at the scene yourself.

- Continuously improve all processes through innovation as well as evolution.
- Dare to pursue perfection, from all details to long-term visions.

To appreciate these five points, one must ruminate on them. Both pillars of the Toyota Way reinforce each other in support of each element. A first-time observer's take of Toyota usually reveals more about the observer than Toyota. A financially driven, results-oriented observer is apt to ask all the wrong questions.

The Toyota Way produces a peopleware system rather than a software system. It is a system for living, not just working. To Westerners, building human skill rather than technology has sometimes seemed overdone within Toyota, especially in Japan, but the overall effect of the Toyota Way has been remarkable.

The Toyota Way descended from the spirit of Sakichi Toyoda, who incorporated Toyota in 1918 to build and market his invention, an automatic loom that stopped whenever a thread broke. Thus Toyota began with a fail-safe idea. Sakichi detested waste. He urged every employee to help eliminate it, but had no systematic means to help them do it. Much later the system that evolved to make Sakichi's philosophy practical was TPS.[4]

THE TOYOTA PRODUCTION SYSTEM

Many people now understand at least a few of the techniques that, to many, have come to represent TPS. There's a lot more to it, but that is where almost everyone starts.

TPS Techniques

If you are familiar with TPS-related tools, skim on; if not, some common ones are as follows:

Kaizen (process improvement): Carefully observe a process in detail. Think of changes that could decrease waste or increase value added. Test changes, including support processes, to see if they work. If so, adhere to the new method as standard. Well done, kaizen is PDCA in action.

PDCA (Plan-Do-Check-Act): A systematic methodology based on the scientific method. Disciplined thinking to identify significant problems as opposed to symptoms; devise countermeasures; test them; and standardize accepted countermeasures in practice.

Visibility: Creates line-of-sight communication among those engaged in a common process. Clues from physical work processes let people know what to do without having to search, ask, or be told. This makes unnecessary much routine verbal or written communication, meetings, and directives. When irregularities are easy to spot, correction begins quicker, so quality improves. Process visibility also prompts kaizen. All other TPS tools help create visibility. *See* 5S.

5S: Everything (material, tools, even computer files) has a set location, and is returned to it immediately after use, ready for next use. A typical series of words denoted by 5S is: Sort, Standardize locations, Scrub (clean), Strictness (following this practice), and Sustain (participation by everyone).

Preventive maintenance (PM): Check-and-correct routines to assure that all equipment works to its full capability whenever needed, as opposed to, "If it ain't broke, don't fix it." Operate by a fire department mentality, with everything ready to go at all times.

Fail-safe: Using simple means, rig a process or machine so that errors are nearly impossible to make. *See* autonomation and *jidoka*.

Autonomation: Design machines to do repetitive tasks; humans always do nonrepetitive thinking. If a defect occurs, or could occur, the machine should stop. A human should check why. Humans think always about the process; machines run only when needed, at the rate needed.

Jidoka: A machine, process, or person stops and makes correction at any sign of defect or error. Never pass on a defect. More broadly, think before working; and in general, think ahead.

Kanban: A signaling system that lets a using operation communicate what it needs to a supply operation; the ideal is use one, make one. Kanban systems often use cards, but exist in great variety. Kanban systems make demand visible between distant points while putting a limit on the inventory between the operations. The real purpose is to make problems visible.

SMED (Single Minute Exchange of Dies): Shorthand for fast change-overs from one type of task to the next. The ideal is instantaneous set-ups to create high variety with minimum waste whenever wanted—processing unique items in lot sizes of one using minimal energy.

Cellular layout: Whenever possible, juxtapose equipment or workstations in the sequence in which work is to be done, avoiding much movement and control of material (or other items) between stations. (Also applies to work done by robots and other intelligent equipment.)

Takt time: *Calculated* time between unit completions based on volume needed. Cycle time is the actual work time at each station. Lead time is the duration of work, start to finish.

Process mapping: A flow chart to create hyperrational visibility of a process that cannot be seen all at once. (Toyota does not term these value stream maps, as in lean manufacturing.)

Objective of TPS: Eliminate Waste

Another basic of TPS is its primary objective: to eliminate waste, defined as anything customers would *not* pay for if they knew about it. A Compression-compatible definition of waste is broader, but the TPS list is a guide to seeing problems, and of course that does little good unless the observers attack the problems—using quality tools or any others. Breaking down waste into the categories below greatly sharpens detection of problems in work processes.

- **Errors, defects, scrap, and rework:** Eliminating this is a subject in itself, entailing a gamut of quality techniques from simple fail-safe methods to analysis of variance in complex cases, plus creating a work culture striving for quality perfection.
- **Waiting (long lead times):** Any material, employee, or customer that is waiting represents an opportunity. Rarely is value added by waiting, as with wood seasoning, or wine aging. If no attention has been paid to process delays, it is not unusual to find that something is merely sitting 95 percent or more of the time that it is in process. Respect for people also demands that processes not waste their time.

- **The processes themselves:** Why waste resources on operations that need not be done at all? Hard-to-make product designs do this. If also hard to ship or to maintain, they waste resources later. The question to ask about administrative procedures and even customer service operations is whether they benefit a customer or are done "just for us."
- **Motion:** Excess motion adds no value (except perhaps in show business). A spaghetti diagram of the travels of materials or administrative approvals frequently produces a tangled tracing suggested by the diagram's name, a waste of time and energy.
- **Transportation:** Related to motion. Unnecessary movement by road, rail, air, or sea wastes fuel, vehicles, roads, and time.
- **Inventory:** Inventory is stuff sitting. Its presence symbolizes delays—waiting—the second waste listed above. Causes are myriad, but a common one is large batch sizes. Paper files, computer files, and even money are also inventory. Managing inventory consumes time, energy, labor, and space. The simplest way to manage it is not to have it. A common saying is, "All inventory is evil," but only inventory that results from unnecessary delay and other wastes is evil. A seed bank or gene pool is not evil.
- **Overproduction:** Working just to be working, for example, because an accounting system requires it in order for statements to look good, or just to look industrious. In general, don't do anything forecast to have value in the future. Instead, develop processes so responsive and reliable that any activity needed can be performed when needed.
- **Information**: Enter it right the first time; keep it simple and useful, and obtainable where needed when needed. Confusing, unwieldy information systems are a huge waste. However, kaizen of information systems can be guided by applying the first seven wastes on this list to nonphysical processes. (See "Visibility" under "TPS Techniques".)
- **Human noninvolvement:** This waste comes from not fully understanding TPS. Those working directly with processes see waste in details that occasional observers miss or never have time to address.

The first seven of these nine bullets on waste correspond to the classic seven that guide TPS. The last two are redundant, implied in the first seven, but sometimes added for emphasis. Nothing on the list is understood without pondering it deeply while trying to apply it; learning to "think process." From physics, one can summarize the elimination of waste as taking any process to its "lowest overall energy state," but the list is a better prompt for things to look for in real work processes.

These tools and wastes—and more—may seem strange at first, but they are not hard to learn superficially. Once onto them, TPS seems to be a commonsense work discipline, but that is on the surface. Learning the techniques starts putting the philosophy into practice. Learning to use TPS to solve problems changes lifetime work habits, but that is still only the beginning. The next insight is discovering that any set of such tools seen is but one example of setting up a system for organizational learning. An infinite variety of "tools" is possible. That is, one tries to structure a great learning process for a specific setting.

Creating a TPS Learning System

Each technique should reinforce all others. When integrated, they structure a system of work that has a system of learning built into it. Implementing one or two tools does not create TPS; it only improves some processes. To have a system of learning, one must involve and develop the learners. Doing so is to *create* TPS, contrasted with implementing it or converting to it. This concurrently restructures work processes and develops rigor and teamwork in the thinking of the people doing the restructuring.

That is, *creating* TPS in this sense changes the work culture. Because work culture is the amalgam of all systems and practices of an organization, creating TPS is much more extensive than using a few industrial engineering tools to improve selected work processes. Tools vary depending on the application, and every work culture is a little different; but in each case, people doing first-line work have to take responsibility for improving it. But for them to do that, managers at all levels have to develop people until they are capable of it. That is harder than it sounds because managers have to abandon vestiges of a master-slave mentality and become teachers or mentors.

Creating TPS is the creation of an organizational self-improvement program. People learn TPS primarily by developing their own TPS-like system. Each physical change, each specific process improvement, deepens understanding. That is, you learn by making changes that help you learn better. It's analogous to progressing through athletic training regimens of increasing rigor until one becomes a world champion.

If TPS techniques are regarded as mere industrial engineering, they result in physical process improvement, but minimal work culture change. Staff engineering can also radically change physical processes, but radically restructuring work processes by creating TPS also radically changes the work culture of a total organization.

Outside Toyota, versions of TPS are frequently called lean production, lean operations, or lean thinking.[5] While the term *lean* suggests organizationwide change, the financially minded easily misinterpret it as cutting costs to the anorexic level, including total payroll and pay rates. Workers fear this, so most companies contemplating lean conversion must assure the workforce that no one will lose a job because of productivity improvement. If they do not, worker fear stalls physical conversion, much less creation of TPS.

However, managers may not foresee how much capacity will be released by eliminating huge amounts of waste, thus underestimating the sales increase needed to fill it up. Without more work coming in, it's impossible to keep a no-layoff promise indefinitely. Eliminating waste eventually cuts costs by resolving many problems previously undetected by the cost system. Thinking through these implications is the key to rethinking business strategies and winning employee hearts and minds. Thought of as a people-development process, lean requires managers to become leaders and mentors of people, which is very different from issuing budgets and directives.

The Role of Standard Work in Kaizen

The heart of TPS is improving work processes by kaizen. All waste is never eliminated, but each improvement opens more visibility into a process, revealing further opportunities for improvement. Small, incremental improvements avoid collapsing into confusion, as when an organization foists a systemwide software change on people unprepared for it. However, small increments add up. Big system changes are required to make them permanent, but psychologically, when process change ideas come from the people doing the work, they try to make them succeed.

To make consistent progress with continuous improvement, TPS must become a system of consistent learning. Otherwise, improvement stagnates by only fixing processes that have been fixed many times before. To avoid this, all improvement should be based on standard work, actually performed as

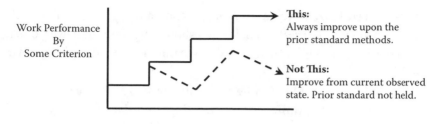

FIGURE 2.1
Effect of improving prior standard work.

documented. The crucial difference is diagrammed in Figure 2.1. Continuous progress is by regular, sequential improvement of standard work.

This is easier to understand by describing it for repetitive factory work. At regular intervals within Toyota, all production employees participate in improving standardized work. The most common interval is monthly, or when the production schedule changes the takt time (run rate), so methods have to change. This is called *distributed production planning*. Changes at the workstation level are much more detailed than just a new production schedule. Starting from basic schedule requirements, employees try out and document new work standards incorporating recent ideas for improvement. This ability allows Toyota's Japanese assembly plants to fine-tune schedules, changing production rates and mixes to match market demand. No Western auto plant has such flexibility.

In most Western plants, the staff develops and documents the work instructions for workers even if the workers devised them by their own kaizen. Afterward, few workers may read the instructions, much less closely follow them. Furthermore, very few Western staff or workers are accomplished in analyzing standard work methods, or in coaching others in how to follow them. To do that, Toyota develops workers to train each other using a four-step training method called Job Instruction (from Training Within Industries, developed in the United States during World War II and then afterward forgotten).[6] Inability of workers to learn work details from others—to share knowledge—is one reason many factories cannot hold standard work and improve on it in stages.

How does Toyota do this? By periodic mini-kaizen on the part of all workers who actually do the work. All help develop it so that all understand it, are able to do it, and as a team integrate the flow of work. Toyota often labels this by the nondescriptive phrase *standardized work*. By contrast, staff-organized kaizen events in Western companies are more like early training in kaizen. To create TPS, leaders should press on until everyone is able to perform kaizen on their own or in teams, as needed, and whenever desired.

Toyota's criteria for redesigning work during kaizen is greatest overall efficiency (not local optimization), evaluating, in order of priority: safety, quality, waste (cost), quantity, flexibility, and visibility. Start by documenting the existing method before trying to improve on it; know why you do whatever you do. Once an existing method is documented as standard, develop new methods to improve upon it. Thereafter, each improved standard method should clearly be better than the prior one by some criteria. Of course, some revisions may be more to function efficiently at some new production rate, or to deal with product design changes. In brief, improvement based on standard work goes like this:

- Observe the current state (method) carefully; document it as the base standard. *Hold* this standard method until a better one is established. The new method may need some fail-safe checks to be sure you are holding it.
- Observe; gather ideas for improvement; ask five whys; do Plan-Do-Check-Act, and so on.
- Design a proposed method. Check it by doing. Document it as operator instructions useful for instructing others in it and holding the new standard.

Toyota has not made any great secret of how this is done, but American managers and supervisors have had difficulty comprehending it. It conflicts with their concept of their responsibility—giving direction.

Keeping people sharp at devising and improving standard work is very important. Without practice, people get rusty. To prompt improvement, even if production requirements do not change, Toyota leaders may change takt time slightly to give people practice. Thus is much process learning built into everyone's work at intervals almost like school terms, rather than being sporadic, management-directed projects. Toyota has those, too, for problems too large or too critical for routine kaizen of standardized work. Management and staff engineer big-step changes that are refined through continuous improvement, but regular, people-powered improvement using standardized work is a big edge. Evidence of detailed improvement and problem-solving activity is obvious in a plant. Work detail at every station is sharply defined, impossible unless the workers create it and sustain it.

Kaizen through standardized work, simple in concept, is disciplined in practice. Furthermore, truly eliminating waste improves processes on multiple criteria at once, without trade-offs; for example, sacrificing quality for speed, or vice versa. Toyota may not use a radar chart as in Figure 2.2 to display the results of improvements, but it illustrates the principle of something (waste) evaporating into nothing. By economic trade-off thinking, such feats should require complex mathematical optimization. It doesn't. Ordinary but well-developed people can use simple tools to eliminate a great deal of waste, if not the most technically complex kind.

If the continuous improvement by kaizen included eliminating environmental wastes, the measurement of many results could theoretically have a common denominator—minimum use of energy (the sum of energy embedded in materials, plus that used in the process). Lack of data limits the practicality of that now.

When used to track improvement, a radar chart such as Figure 2.2 (it's hypothetical for illustration) should show improvement in at least one dimension of performance without a decrement in any others—no

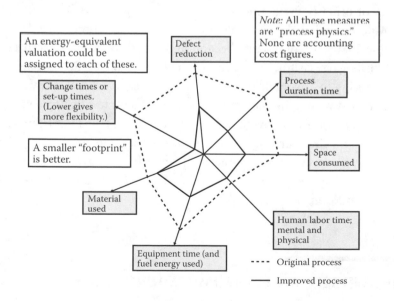

FIGURE 2.2
Process compression by multiple criteria using kaizen.

trade-off. Of course, process conditions and requirements change, as when taking on a more complex product design, so not every change in standard work will be an obvious improvement over the last. But over time, continuous improvement should be a progression of improvements over the original.

Improvement is never done. Before a process can be pared down to its waste-free nubs, technology changes, new products are introduced, or market demand shifts. The notion of coasting on a best method for years is "so twentieth century." In Compression, every working organization needs to embed very efficient learning processes in its DNA.

Transform factories and other physical operations into learning laboratories, with products and processes easy to see. Toyota does. Every work area is a learning laboratory for coaching people in process improvement, process innovation, and maybe in experimenting to make learning processes themselves more effective.

Almost all new Toyota staff and management begin with a stint in a factory, learning basic TPS, how to support it, and how to apply its thinking in nonfactory contexts. In the 1970s, Toyota plants became learning labs for suppliers learning to create TPS inside their own companies. To gain a good understanding of TPS (or lean operations), every company needs a factory or other physical operations as a learning laboratory. Only learning by doing makes the principles of the Toyota Way come alive.

In the 1980s, Toyota plants became industrial tourism stops in Japan. Brief visits gave visitors a misleading surface view of TPS. Extension of TPS thinking to Toyota dealers has never been satisfactory. Many of them have dabbled with it, but being legally independent, few dealers ever work a stint in a plant. Even Toyota, where TPS originated, has not been able to imbue its TPS-like learning processes end-to-end in its total supply chain. Internally, however, TPS as a practical primer in the Toyota Way has dramatically influenced every system and every work process, especially new product development.

___

THE SEEDBED: AN ECONOMIC MICROCOSM

The seedbed from which TPS emerged was a working community isolated from the economic mainstream. In the 1950s, Toyota and all its primary suppliers were in Aichi Prefecture, little bigger than a midwestern U.S. county. This automotive enterprise was probably more geographically compact than any other in the world, before or since, but in dire straits: 10,000 people produced fewer than 25,000 vehicles per year; capital near zero; dismally low productivity; no technology to sell, and no way to buy any.[7] Toyota was always 10 yen from disaster. Motivated as if fighting a war, Toyota's TPS pioneers strove to outperform all competitors by superior use of basics.

In the 1950s, Toyota's survival was crucial to the Aichi community; they sorely needed hard foreign currency to stay alive financially. Besides developing all the people, circumstances dictated four other objectives:

- Promote a community spirit (suppliers, the people of Aichi, and so on).
- Spend no more than one-tenth of the capital per vehicle built than competitors.
- Make a plant break even at 30 percent of capacity (implies a high margin).
- Create the effect of automation (do-it-yourself; autonomation; and so on) without buying anything expensive.

Sometimes Westerners think that TPS is an adjunct of Japanese national culture; Japanese don't. National culture may have influenced TPS, but many, many other Japanese companies were also desperate at the time. Only one fully developed TPS: Toyota. A better explanation is that not only was Toyota nearly down and out, but it also was isolated in Aichi Prefecture. Not knowing much about industrial or business conventions, they were free to risk doing anything that made sense. All the key players knew each other, and could easily meet to try out new ideas. Although Aichi was relatively rural, most Toyota workers weren't fresh off the farm, but from all over Japan. And they were young, free of old legacy thinking. Housing being scarce, dormitories were a big attraction. Working hard and playing hard together, they melded into a strong-bond community.

In this economic microcosm, face-to-face meetings were frequent, communication was easy. Experimenters mixed and modified sketchy ideas that scouts brought in from elsewhere. With little outside pressure to conform to other companies' expectations or systems, and unaware of industrial conventions, they were free to innovate however they saw fit.

The system Toyota created in this learning laboratory may have already become as influential as Ford's assembly line earlier in the twentieth century.

Competing with mass production giants, TPS largely offset the effects of economy of scale (something lean enthusiasts may not know—or have overlooked). TPS, and later the Toyota Way, is an organizational innovation. They contradict the business orthodoxy of direct control to assure a profit. Successful creation of a learning organization through TPS or the Toyota Way requires leadership of the people, not management for the capital.

OHNO'S METHOD

Old hands at Toyota often refer to TPS as "Ohno's Method," after Taiichi Ohno (1912–1990), frequently cited as the father of TPS. He took no credit for inventing any of its techniques himself (with the possible exception of kanban). Ohno pushed others to invent and innovate, and he was not Mr. Charming doing it. He challenged people to tackle problems they would have preferred to ignore, and has been likened to Mark Twain's Tom Sawyer snookering his friends into painting the fence. A sharp observer, Ohno preferred to investigate a manufacturing problem at the scene of the crime, not through abstract reports. He coached his budding TPS leaders to acquire the same trait, turning his wrath on those torpid in observation or unimaginative in thought.

Ohno asked tough, probing questions of himself and others—why, why, why, why, why—at least five times, not stopping until the root cause of a problem seemed evident, so that something more than a patch could be put on it. No jumping to conclusions, no muddled logic, and no wimpy attitudes about what isn't known or can't be done.

One of Ohno's methods is called "standing in the circle," in a factory, office, or anywhere work processes can be seen. Drawing a chalk circle on the floor, he would tell a budding TPS leader to stand in it and watch. Don't break concentration; don't run off to do something that seems important. Absorb a complete picture of the scene, seeing processes as a whole—and in detail: layout, machines, tools, material, movement patterns, human issues, paperwork, knowledge gaps, and on and on. Don't grab for a solution without understanding a problem in full context. (This instills discipline to absorb facts from the scene yourself, now part of the Toyota Way.)

Ohno's cadets did not literally stay in a circle, gawking like tourists. They probed until they grasped a full picture of the facts at the source (*genchi gembetsu* in Japanese). Quick fixes were unacceptable; no heroic celebration; keep going. If a machine has a faulty bearing, don't just replace it. Why did it fail: misuse, negligent maintenance, poor design, or something else? Run each problem down to its roots, and grub out the roots.

Think ahead and think afar. Avoid crises and create opportunities. Force yourself to solve problems without spending money. Coach others to do likewise.

Ohno and his disciples were strict *teachers of process improvement.* Doers of work had to constantly question everything they did, and continuously improve, devising their own work methods. They had to understand why, not blindly follow directions.

This kind of leadership may be impossible for so-called control freaks, who think that their position mandates that they direct people overtly—or subtly by using cost controls and systems menus. TPS leaders demanded that doers think for themselves and take real responsibility—and develop the expertise to exercise it. The leaders' first responsibility was to develop people to do this. Development of others, not business results or personal accomplishments, is the top item assessing a mentor-leader's performance. To this day, claiming excess personal credit is behavior that will be ostracized by other TPS leaders.[8]

People so biased for action that they must constantly make something happen can't stand Ohno's method. Veteran TPS leaders will "stand in a circle" only those candidates who prove patient and interested. They have to come to the Toyota Way; it will not come to them.

TPS is the opposite of permissive management. It is far more disciplined than the illusion of discipline that a boss has when giving people orders. Ohno was demanding. He insisted that people solve problems, collaborate, and take responsibility. Doing their best was not good enough if better was possible. A gruff shopman, Ohno was a complex combination of inquisitor and irritant to the status quo, definitely not a deft corporate politician. Had the Toyota family not supported his methods, TPS would probably never have happened.

But TPS mentoring, however tough, really is respect for people: respect for who they are and what they can do when developed to function at full potential. A turning point leading to this came in 1952. Toyota's first and only strike, triggered by a layoff, rang an alarm. Unless they survived aboard the Toyota ark, workers would not enthusiastically contribute to process improvement. From that time onward, employees were respected as people, regarded as the heart of the company, not as commodity labor.

Almost all of Ohno's original cadets have retired from Toyota. Some continue working outside Toyota. Second- and third-generation TPS leaders modeling their behavior after Ohno can trace their mentors in a line back to him.

TOYOTA'S SUPPLIER ASSOCIATION

Many manufacturing companies everywhere sponsor supplier councils and have supplier meetings. These fulfill various needs—feedback to the customer, socializing, technical changes, performance needed, recognition, or just an opportunity for the customer to announce new cost-reduction targets to the suppliers.

Supplier associations are a long Japanese business tradition. Many are just social clubs among business friends, but Toyota's supplier association was unusually rigorous. At times it was a boot camp to teach suppliers process improvement. In the 1970s, all Toyota suppliers in Japan belonged, except vendors of commodities not critical to quality.[9]

The OPEC oil embargo of 1973 jolted Toyota's supplier association. To create TPS in supplier companies, the Toyota community around Aichi went back into survival mode—create TPS or die. Ohno's TPS leaders fanned out through supplier plants.

An example of the intensity of this campaign was Tokai Rika, which supplied almost everything in an instrument panel. The first try utterly failed. Senior management, like Toyota in 1952, realized that they had to become one with the workers—personally lead them. Calling everyone together, with no speechifying, they began to sweep floors and wipe machines, symbolizing that we're all in this together to do whatever it takes.

That was the turning point. By day they made parts; by night they made changes. Five months later, Tokai Rika's Otowa Plant had converted to one-piece flow. One of the top managers said that during that time he had gone home only three times. Going through this upheaval on the floor, with workers, changed Tokai Rika's work culture.[10]

With the conversion of primary suppliers to TPS, the Toyota community not only became the center of a system of production that others are still emulating, but *it also developed the basic concepts of supply chain management*—when the Internet was still a dream. The key was cross-company collaboration by operating people. It still is, at any level of technology.

People routinely went between companies on assignment, as visiting engineers for example. Toyota and its suppliers assigned promising young people to kaizen teams on round-robin tours, doing a project at each company. The short-term benefit of these kaizen tours was limited, but the long-term benefit was priceless. Years later, during conflict or crisis, key people in the supplier network knew each other very well.

For example, in 1997, an Aisin Seiki single-source plant was nearly destroyed by fire. Scoffers thought the catastrophe would spell the end of TPS just-in-time delivery. Instead, the supplier network went into fire-drill mode. Tools were rescued, refurbished, and installed with other machines in sister plants. Production returned to normal within a week, almost as if a plan for that exact contingency had been rehearsed.

Suppliers routinely collaborate in product design. Keeping up technically is obligatory; technical obsolescence is cause to be discontinued from the community. Indeed, being a Toyota supplier is analogous to being on a top sports team; staying in condition for top performance is a minimal expectation. By today's standards, one needs to be an aggressive learning organization to do it. New product development processes at key suppliers are similar to that used by Toyota. Suppliers offer the best alternatives they can devise to meet the design challenge of a vehicle—a little like hackers collaborating on open software.

The relationships between Toyota and its key suppliers entail much greater performance discipline than mere contracts can describe. A code phrase for it is "coexistence and mutual prosperity." That means, "If you die, I die with you," and "If I manage to gain a little, you can have part of it." The ideal is work organizations that are neither so starved that they are weak, nor so fat that they are complacent, but always ready for peak performance.

THE TOYOTA PRODUCT DEVELOPMENT SYSTEM

The learning structure of the TNPD system differs from that of TPS, but it diverges from business orthodoxy by following the same Toyota Way philosophy. Making very few errors, it quickly and efficiently designs highly reliable vehicles with bland styling. Primarily it is a technical knowledge learning system. Secondarily, it develops new products.

Most Western companies structure a new product development project to research, design, test, and launch a specific product, or a family of them. New technology is often developed as part of a new product project. By contrast, Toyota engineers develop little new technology specifically for a new product (the hybrid-drive Prius was an exception). Instead, they test and codify their general knowledge of automotive technology. From this knowledge bank, they select packages to integrate into a new vehicle design as late as possible. This is called *set-based design*.

Western companies tend to document their technical knowledge in files of projects from which it originated. A team starting a new product project may not find old project files easily, so Western companies are developing more accessible knowledge management systems to codify what they know.[11] But their technology and product development personnel are not as likely to use the system in daily work procedures as Toyota, where all chunks of learning have to squeeze onto one sheet of paper, called an A3 paper. Regularly making this mental squeeze forms habits of clear thinking, and retrieving knowledge is much quicker and easier if it was clearly described when filed.

Michael Kennedy calls the Western approach "design, and then test," in contrast with Toyota's "test, and then design."[12] Toyota emphasizes building up a usable, accessible knowledge base of findings about specific technical problems or issues, including trade-off curves that relate physical parameters to cost. Designing a vehicle to sell at a target price-point requires translating technical characteristics into current costs, then integrating a package of known technology to best fit the needs of prospective customers for a price they will pay. In effect, costs are regarded as just another design spec.

Toyota develops new vehicles in roughly half the time of Western competitors, using perhaps one-fourth the engineering hours, and almost always on time, within budget, and with fewer design flaws that lead to recalls. Western new-model projects are well-known for being overdue and over budget.

The system outlined in Figure 2.3 is called the *Shusa* (chief engineer) System. The final design is not by consensus, but the responsibility of the chief engineer, although many negotiations may take place determining

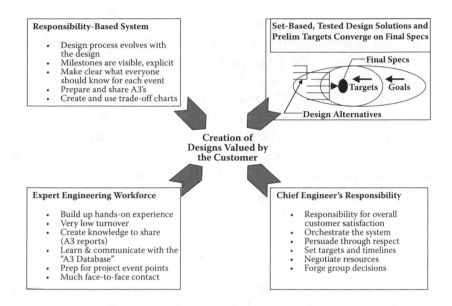

FIGURE 2.3

Overview of Toyota's new product development system. Figure based on a presentation slide by Michael J. Kennedy. With permission.

how to design for various requirements, from fabricating tools to field maintenance. Before starting design, the chief engineer spends six months or so personally seeing drivers and driving conditions, checking the needs of intended customers firsthand with a design eye. In addition to marketing data, the chief engineer develops a personal vision of the vehicle. For design work, the chief engineer's dedicated project team is small; he has to draw resources from technical departments. Design integration centers on a war room called an *obeya,* where reverse-engineering displays, prototypes, and data quickly convey project status to anyone coming in. This system records data on what didn't work as well as what did (to keep from reinventing the wheel), and sustains a consistent product vision.

The Western approach to learning is deductive, like Western science: Set up logical hypotheses; test them by experimentation. Following this pattern, Western new-product engineering creates specs intended to meet customer needs, designs to those specs, then tests to validate the design. If a test fails, redesign of one part of an integrated system may loop back to force redesign in many other parts. These loop backs chew up time—a sinkhole for waste—while designs are becoming more and more tightly integrated.

Japanese learn more inductively, by observing phenomena and integrating conclusions from what is actually seen. Using this inductive approach, Toyota precludes many design loop backs. They continuously observe and document technologies of interest, including those for production processes, and documenting those that don't work, and why, as well as those that do. For example, to understand body fit and finish, study the deflection of various materials when they are clamped in fixtures and heat warped by welding. Organize, codify, and compare findings. Summarize these in trade-off curves separating feasible and infeasible zones. To design a vehicle body, evaluate alternatives using this data. (Western engineers do this, too, but seldom have compiled historical data as comprehensively as Toyota.)

Expert Engineering Workforce

Toyota concentrates on developing deeply experienced engineers. Most stay for a career. To keep from going stale, most rotate assignments, sometimes investigating new technology, sometimes in sales showrooms, and of course, in factories. Manufacturing is a customer of design, too. Toyota wants engineers to be T-shaped, technically deep, but broad on top, the better to contribute to integrative problem solving. And very significant is that being an expert is not highly prized unless the expert helps others to extend their knowledge. That is, there is no glory to gain by keeping knowledge secret.

Responsibility-Based System

All engineers learn how to read and prepare A3 papers. Until recently they were not in computer databases, but papers in files. To see them, one had to go to the source, with opportunity to ask questions of a source engineer or department. A vehicle design project may not be scheduled in detail at the beginning, but meetings are announced in advance, and everyone must come with alternatives in mind. The system works because engineers learn to work within this process discipline.

Set-Based Engineering

This refers to selecting among alternatives late in design. It works because a design is not immediately reduced to geometry with sub-system envelopes to fill. Alternatives may include more than one set of basic concepts to meet the same objective. Several prototypes may be mocked up to visualize alternative sets before a final selection is made. Production people, maintenance people, and others review these for practicality from their view. (Western versions of this are often called 3P—Production Preparation Process.) In the Toyota version, specifications may change as different concepts are considered, but tighten as design jells in a sequence of decisions.

The Chief Engineer

The chief engineer's months of pre-research on customers must result in a technical vision (with cost targets) that can be imparted to every participant in design, including supplier engineers. He is charged with all final design decisions, or with creating a consensus on them. Vehicle design and cost is his baby. Only a creative senior engineer with the respect of peers can do this job.

The heart of the system is A3 papers and trade-off curves—an engineering version of standardized work—complemented by being able to contact the originator of most papers. A system so heavily dependent on interaction between people worked remarkably well in Aichi Prefecture, but it is not easy to transfer around the globe. Growth is forcing Toyota to deal with that.

ETHNIC CULTURE AND THE TOYOTA WAY

Few Japanese familiar with Toyota think that the Toyota Way is strongly dependent on Japanese culture. Evidence on this is mixed. On one hand, many Japanese companies, like Western ones, implemented TPS techniques without understanding the Toyota Way. On the other hand, even the name "the Toyota Way" rings of oriental philosophy. Japanese may not be the best judge of this. Like the rest of us, they can't see their own culture clearly.

Japanese culture, like all others, is self-contradictory, ever changing, and now overlaid with Western influences. The feudal Samurai system, depicted in movies, long ago gave way to Japan's distinctive harmony for mutual survival on earthquake-prone islands. Pre-war Japanese work culture seemed much like work cultures in Europe.

None of Japan's collaborative traditions, like rice farmers sharing irrigation water, were consciously drawn upon to develop TPS or other Toyota systems. Culture was more like background radiation. For example, Japan's traditional sports—sumo, judo, archery—are as much about deeply pursuing as competitively winning. Adding to confusion, Japanese often relabel what they do to make it interpretable to Western business.

For example, *keiretsu* now means a collaborative operational supplier family or supplier association, but the original postwar keiretsu were formed to bootstrap economic growth despite capital shortages. Companies in each keiretsu helped each other while a central mother bank nursed them all along, and if the bank was short, companies helped fund each other (although cross-holding of stock is a custom that predates World War II). Companies in the same keiretsu sometimes teamed up to take on a large contract.

The last big keiretsu project was in 1974, but Westerners thought keiretsu family financial ties were a precondition for operational collaboration between Toyota and its suppliers.[13] Not so; operational collaboration always depended on relations between the people in the companies, regardless of financial ties. But because of this misperception, Japanese themselves began to term collaborative supplier groups as keiretsu. After Japan's financial bubble burst in 1991 and Western companies began supply chain management, most Western illusions of keiretsu financial magic faded away.

Close operating relations with suppliers, as well as TPS, grew out of experiments by operative personnel. However, measures to promote mutual financial security did have an indirect effect: they fostered lifetime employment as an unofficial custom. That made investing in long-term development of people a good bet, one on which TPS came to depend. Consequently, many Japanese boards are still more likely than Western ones to regard a company as its people and whose mission is to serve all its stakeholders, not as an entity to generate maximum return to owners. On the other hand, Japanese youth today have much less affinity for lifetime employment than their fathers, so companies have to work harder to assimilate a loyal cadre into the core of their working cultures.[14]

In both Toyota and its key suppliers, trust, common goals, and a common basis for problem solving are essential for collective experimentation. Such openness and visibility is the opposite of secretive financial control—by crony capitalists or anyone else. It forms the basis for a learning culture based on *rigorous learning methodologies* and *interactive behavioral skills*. Such things are not intrinsic to Japanese ethnic culture. Japanese also have financial scandals and their own mafia (*yakuza*).[15] But a learning work culture is difficult to understand by any business leader whose concept of a work organization is limited to being an earnings asset of a financially tradable entity.

However, TPS was born in a high-context ethnic culture. Japanese have the highest context culture of any industrial society. Edward T. Hall originated the concept of high-context versus low-context cultures, shown in Figure 2.4.[16] Cultures may differ substantially, depending on the degree to which they are based on relationships (high-context) versus structures (low-context). High-context cultures are more aware of relationships between people and between things around them. Low-context cultures focus on things and events. Low-context people see objects; high-context people see the space between objects. Other observable differences are shown in Figure 2.4. (An extremely high-context tribe may exhibit almost zero concept of time, for instance). In general, Asians are higher-context than Westerners.

Some experiments with high- and low-context cultural differences asked people to describe an aquarium. American eyeballs immediately glommed onto the biggest, brightest, or fastest fish. Asians tended to describe

Factor ↓	Relationship Oriented (High-Context)	Structural Oriented (Low-Context)
Personal Time Concept	Polychronic–multitrack Tolerate interruptions *Relationship driven*	Monochronic—single track Punctual, focused *Goal and control driven*
Space Concepts	Integrative Boss sits with a group Use dense, holistic graphics	Allocated, segmented Boss has private office Prefer simple graphics
Problem Solving	Inductive	Deductive
Division of Work	Interactive responsibilities	Assigned responsibilities
Method of Social Change	Adaptation	Fractionate, recombine
Moral Driver	Group acceptance (shame)	Personal integrity (guilt)
Types of Self-Contradictions	Strict process schedules	Excuses for being late

FIGURE 2.4
Comparison of relationship and structural cultures. Based on the work of Edward T. Hall.

context first, the frame or background of the aquarium, and then the fish. In another experiment, Americans demonstrated better spatial alignment perception when they thought (incorrectly) that they controlled the alignment; Asian perception showed no such control bias.[17]

High-context people are expected to know what's going on about them. They communicate more for that purpose. Action messages may be short, perhaps implied, like a slight nod. Low-context organizational action messages must spell out more detail. The epitome of this is a detailed legal contract. A high-context person, relying on personal trust, may see no point in having a contract.

One can only speculate beyond the research, but doing so is intriguing. Are high-context people better able to see what should be in a picture, but isn't there? That ability is important to learning.

A low-context Western work culture may have more difficulty instilling the patience for detailed observation, visible processes, and careful learning, sought when TPS is created rather than installed. Low-context, structurally oriented organizations talk about poor communication, but can't put their finger on why. They tend to focus on results, deals, transactions, and financial assets. Process relationships in a broader context may escape them.

Despite their high-context ethnic culture, Japanese fell into "business think," too. Complacency is one of its blind spots, just as in the West. If profits represent success, concentrating on them produces euphoria followed by inattention (bubble and pop). For example, Nissan paid the price for falling into complacency with its suppliers.[18]

The Toyota Way tries to inoculate people against complacency and arrogance.[19] It concentrates on what we do and on what we can do better, not on what rewards we may get. This turns the business status system upside down. It conflicts with normal investor motivation and clashes with global corporatism. Even today, no matter how high profits may be, Toyota never officially boasts or draws attention to them, but emphasizes the enormity of its future challenges. That attitude is necessary to meet the challenges of Compression, something that not even Toyota has yet begun to do.

Lean Manufacturing

In the early 1980s, American companies began implementing knockoffs of selected TPS techniques. Some merely grafted one-tool fads, like Quality

Circles, onto nonreceptive working cultures. These wilted like orchids in the Sahara. A few companies like Hewlett-Packard and Harley-Davidson implemented more extensive suites of TPS techniques, but still fell far short of creating a working culture resembling TPS.

Intensive kaizen events that began at Jacobs Vehicle Systems (Jake Brake) soon became part of American TPS-like systems. In 1990, *The Machine That Changed the World* coined a name for such systems—lean manufacturing.

But American companies had had to learn quality improvement quickly. For example, increasing complexity ballooned the opportunities for defects in circuit boards. Without better quality processes, repairing defective boards would be never-ending. And to survive, American auto quality had to approach the Japanese level. To be an auto supplier, quality certification became necessary. By 2007, J.D. Power ratings of Western mid-class vehicles had considerably narrowed the gap with Toyota and Honda.

Little by little, American manufacturers learned that only the improvement of basic process physics reduces waste, improves quality, increases flexibility, and decreases lead times. If the basic process is poor, automation and complex scheduling have relatively little impact.

By 2007 a clearly American pattern for lean manufacturing had emerged. Skipping lots of details, it goes like this: Map the flow of a process end-to-end (called value stream mapping so the financially minded can identify with it). Initiate 5S and visibility systems, making flow problems easier to see. Align operations for one value stream physically in one area (cells or focused factories). Organize periodic kaizen events to attack the primary areas seen to have waste. Involve the workers in these events; cross-train them for new and multiple tasks. Reduce set-up times. Repeat. And repeat. Then extend these initiatives to offices, shipping and receiving, product design, suppliers, and sometimes dealers or customers.

If sustained, these changes often reduce a remarkable amount of waste. Defect rates, lead times, and similar measures improve by two to ten times, or even more, which seems astounding. However, much of that waste had accumulated over time because the prior system had not been designed for anyone to see it and take serious action. That, too, stems from conventional business bias. An investment-based concept is to invest in an asset, then wring the highest return for ownership from that asset, investing in things, like software, more than in people and learning processes.

Consequently, Americans rarely see lean manufacturing as fully developing a workforce. Rather, they reprogram the work culture just enough to make the techniques work. For example, they coach people to work in teams, but don't stimulate them to always be learning on the job. And many have misconceptions, believing that lean thinking is a set of tools only for volume production, or shop floors, or cost reduction. It is not seen as a fundamentally different philosophy of business, a big mental stretch.

Those who define lean as its tools frequently install hybrid techniques. Lean Six Sigma is a favorite. Some think that the more tools in the kit, the better, rather than the skill and consistency with which they are used. As agents of ownership, most drive lean conversion toward desired results using key measurements often called dashboards.

Following their Taylorist legacy, tool-minded managers engineer lean process methods into place, including any supporting software. In the worst cases, they set up an adjunct department in their bureaucracy to implement lean. Typical problems are as follows:[20]

- Managers implement techniques, but minimally develop people to take initiative overcoming process problems. That is, they don't fathom TPS as a medium for stimulating a learning system—no version of standardized work, for instance. Workers mostly run a daily routine until directed to the next kaizen event.
- Managers assume that lean is a cost-cutting program so that conventional cost systems should quickly show results. If they don't, they are tempted to kill the program. But even after adopting "lean accounting," they may maintain reward systems that drive people to meet results goals. Letting go of controls and directives, developing people to improve processes themselves, is fearful. (Ownership is apt to blame its managerial agents, presumed to be "in control," for any shortfalls.)
- Managers protect too many old institutional monuments—piece-rate pay, legacy software, bureaucratic interpretations of regulations like Sarbanes-Oxley. When slow at removing such roadblocks, process simplification slows to a crawl or stops.

After big initial reductions in waste, process improvements are not as startling, so management assumes that "the system" is installed; no further cultural change is necessary to sustain the gain. That's an unstable situation. Going on to develop an ongoing process learning system will not only prevent huge amounts of waste from accumulating again, it also eliminates a lot more waste still unseen at that point.

If such a system falls back under financially driven business leadership, process changes can be quickly undone. Creating anything like real TPS—with a learning system—is a massive change from a transactional mindset. Just acclimating people to a discipline like standardized work may take three years or so. Leading such a human transformation is no small challenge, even for those who understand what they are going for and why.

Despite these problems, lean thinking has made a difference. By 2002, all ten of *IndustryWeek*'s America's Best Plant winners showed evidence of TPS-like practices.[21] The Shingo Prize annually honors leading lean plants.[22] Companies are using lean thinking to streamline nonmanufacturing processes. And nonmanufacturing operations—like hospitals and law enforcement agencies—have begun pioneering lean thinking.

In service businesses, concepts from manufacturing seem out of place. For better acceptance, service industry conversions may prefer to label process improvement with a name like "quality," for instance. However, the basic principles are the same, the basic issues very similar, and without developing a learning work culture, the revolution stalls there, too.

Learning and Work Culture

Unless a work culture has a built-in learning process, technology and technique changes do not assure superior ongoing operational performance. Overconfidence in techniques and technology leads to complacency, somewhat like drivers learning to use antilock brakes. By applying them later and later from a higher speed, they negated the initial safety improvement, an effect called *target risk*. Performing at a risk level with which a person feels confident may be a human instinct.[23] If so, any system to overcome this human instinct for complacency has to offset degradation to the target risk level.

A classic case of complacency is NASA and fuel tank insulating foam. In 2003 the *Columbia* space shuttle broke up on reentry into the earth's atmosphere. A chunk of foam had broken off during launch, puncturing a wing heat-shield panel, and reentry heat brought the shuttle down. Foam had broken off on many prior launches. This was worrisome, but not a show-stopper because launching on schedule was high priority. NASA's can-do management came to assume that staff should prove that the shuttle *isn't* ready to fly, rather than prove that it *is* ready to fly, a huge difference in burden of proof. Foam break-off was a tougher problem than acknowledged and therefore easily shrugged off. No test simulations were deliberately biased, but interpretations of them favored rosy assumptions and ignored nagging doubts. The rationale for complacency is that we're doing the very best anyone can, a psychology not easy to overcome.[24]

After grounding shuttles for months, risks from flying chunks of foam were considered acceptable. But on the very next launch, a 0.9-pound hunk of foam broke from a part of the fuel tank that had never before shed foam, the *Discovery*. This time it was not a disaster, but it grounded shuttles again. The commission investigating this was the first to cite an organization's "working culture" as a key factor in a major disaster. It concluded that NASA's work culture had regressed into complacency similar to that preceding the 1986 *Challenger* disaster.[25]

But why? Two major reasons stand out. First, quantity (number of launches) trumped quality. Second, the process-learning system built into NASA's culture decayed until it was inadequate for the complexity of its challenges—assuming that at one time it had been adequate. A built-in learning system means that discovering, codifying, and reusing knowledge is standard procedure associated with all work processes. Because system structure, a perfectionist attitude, and skill using a learning system are all necessary for it to be rigorously used, fixing one that's broken takes more than a management mandate.

This example was selected rather than one from business because in theory NASA is a nonprofit. Because the shuttle system is old, it's easy to lose that first-time edge of adrenalin, while the technology remains complex. Wagging fingers at political decisions is tempting: The foam problem can be linked to limiting NASA's budget, experienced staff leaving the agency, NASA contractor personnel being rushed, and on and on. But all these add up to that old nonsense cliché about accidents: "Had we known what was going to happen, we could have prevented it."

In a nutshell, if NASA can learn anything from the Toyota Way, it's to always be learning-driven instead of results-driven. Lots of TPS or TNPD techniques might apply, too, although they must be adapted, but the key is to never compromise a culture of perfection by drifting into complacency. No matter how much market it captured, or how much profit Toyota made, public statements list accomplishments matter-of-factly, without fanfare, and always emphasize what more must be done.

Common Obstacles to Success Implementing Lean Operations	Elements of the Toyota Way
Can't break departmental silos or open minds to a larger view of responsibilities	Learn to see integrated, holistic processes extending outside the company
Managerial urge to control Ownership interests dominate	Fully develop all people; share knowledge and opportunities
Communicate via chain of command	Create visibility for rapid, decentralized operational communication
Keep old finance, accounting, and IT models Keep old performance measures	Create "organizational learning" based on facts and evidence "at the source"
Instability of leadership and direction by pursing short-term goals	Never stop going for excellence, both in detailed activity and in attaining long-term visions

FIGURE 2.5
Working culture differences.

The Toyota Way is also tough for Toyota to live up to. But for decades, it has done it better than most well-known companies, few of whom even try. That is a major reason why, by expansionist standards, the company's performance is a global benchmark.

Figure 2.5 summarizes of 25 years of anecdotal comments heard from lean change agents bewailing why they and their organizations are not progressing with a lean conversion.

From a working culture view, why do most companies fall short trying to emulate Toyota? The reasons for the contrasts listed in Figure 2.5 lie buried deep in working culture DNA—and old business legacies.

Company leaders attempting to emulate Toyota generally cannot see the cultural changes that are associated with learning systems, like TPS techniques, until they begin implementing the techniques and bump into human recalcitrance. For example, drawing value stream maps begins to open minds, the first issue in Figure 2.5, but only the minds of those who drew the maps, not everyone.

This problem stems from regarding learning to see as part of a sporadic, physical process-change event, rather than as a habit of thinking to inculcate in everyone. Without cultivating this habit of mind, progress stagnates because nonobservers tend to:

- Remain ignorant of processes remote from those they see—other departments, at suppliers, at customers, in nature, and so on

- Never completely see those processes that they do observe; that is, they've never done the equivalent of Ohno's circle exercises
- Never see how their own behavior affects others, which is probably the hardest kind of observation to make

The third point greatly affects daily communication, much less problem solving. For example, people trying to win in meetings don't listen except to spot their next opening, adeptly cutting off other people at the right moment. (The behavioral jargon for this is "inability to dialogue.") Efficient collective problem solving is as much a matter of behavior as clear, logical methodology. Both are necessary.

Behavior is changed primarily by learning how to behave differently. To change behavior, change the environment that provokes the behavior. All lean and quality techniques imply behaviors associated with them. For example, creating a visibility system requires people to trust one another. For example, suppose technicians have always owned their own tools, keeping them in their own toolbox. Spreading all these out on commonly accessible shadow boards is a big jolt to the social system, not absorbed quickly or smoothly. Everyone—or the company—has to literally own the tools jointly. And that is a minimalist change just to get started.

Learning to dialogue in team meetings, respecting others, is a behavioral modification. Learning to carefully observe work so that one can train others is a behavioral modification. Learning to actually use a structured problem-solving approach is a behavioral modification. And learning to lead people into modifying their behavior to participate in rigorous learning processes is itself a behavioral revolution for managers who must learn to lead in a new way. Being aware that behavior must change helps, but talking about better behavior without actually changing the work environment that prompts it does little good. Learning to objectively observe processes and improve them is also a behavioral modification.

For people working in any capacity to adopt a professional attitude, taking on as much responsibility as possible, these kinds of behavioral changes are fundamental. Work organization leaders have to lead by example, illustrating the new behavior expected of everyone. And they must not only hand the responsibility for work to the people doing it, but help them develop technically and emotionally to assume it.

That means that discipline like a parent–child directive model must modify into a professional learning model. For example, improve a work process before disciplining anyone (writing them up) for not following it. (Egregious neglect is an exception.) Developing people to fully engage in standardized work shows respect for them and accomplishes far more than berating them. Nonetheless, not everyone is cut out to work in this way. Over time, peer pressure will likely remove those who can't from the workforce. But the experience of lean companies is that managers and supervisors have more difficulty abandoning parent mind-sets than workers have abandoning child mind-sets.

None of this is helped by retaining conventional business measurements. Financial statements presume separate companies independently contending for growth and profit. Inside companies, cost systems may further balkanize departments using separate, often conflicting performance measures (and bonus systems). Transaction systems connecting them are audited. Such a system looks logical; everything adds up in spreadsheets. "In control" means that performance meets expectations. If it doesn't, is it because performance was poor, the expectations poor, or both?

Inside one of these controlled entities, participants concentrate on local concerns and rewards. They cannot concentrate on process improvement and bigger pictures as long as measurement systems discourage it. Another significant problem is that accrual accounting shunts overhead allocations in strange ways, does not quickly reflect physical changes like decreased inventories, or even shows that decreasing waste costs more—usually because reduced levels of activity underabsorb overhead. That is, accrual accounting is a time warp removed from real physical operations. In recent years, a reform movement known as *lean accounting* has begun.[26]

With few exceptions, *lean* is staff-installed with a minimum of human development.[27] However, merely installing TPS techniques makes startling improvements in lead times, inventories, quality defects, and productivity—if they can be sustained. Learning to progress beyond this is, for most of us, a gradual eye-opening experience. It leads to the conclusion that a work organization cannot be regarded primarily as an ownership's money-making machine, but has to be regarded as a professional organization in the service of all its stakeholders. This is counterintuitive to the Western commercial mind-set because deep differences in thinking penetrate into every crevice of management thinking. Some are outlined in Figure 2.6, which could have run on for pages, but readers are invited to ponder the differences for themselves.

Financially Driven Business	The Toyota Way
Profit and growth are the primary goals.	Survival is primary; profits let you survive.
A company is primarily a financial entity.	A company is primarily people.
Develop a few people as leaders. Train nonmanagement workers.	Regard *all* employees as professional. Develop all of them to their maximum capability.
Leaders design and direct work processes.	Leaders primarily develop people.
Managerial status depends on results.	Managerial status depends on developing people.
Quality is a trade-off.	Quality always comes first (after safety).
Success is monetary. Enjoy success; milk a cash cow.	Never become complacent, accepting the status quo. Always strive for perfection.
Make money from assets: business models, brand images, patents, software, etc.	For "capital avoidance," develop people to "do-it-yourself." Ingenuity before investment.
Top-down communication dominates.	Much horizontal communication expected.
Most action is directed; supervisors or managers deliver many action messages.	Many action messages from the process itself. Routine operations "run themselves."
Motivate people with financial goals and performance measures.	Develop people to improve processes. Lead their development by mentoring.
Assume employee turnover. Hire talent as needed. If profitability is threatened, all employees are vulnerable to dismissal.	Assumption of low-turnover employees. Try to keep margins high to preclude dismissals in downturns.
All process changes should be cost-justified by an accounting model.	Make physical processes broadly visible. Improve physical performance.
Limit access to data without a need to know.	Most company data is open.
Guide improvement by financial thinking.	Guide improvement by scientific thinking applied to real processes (like PDCA).
Improve my personal work or improve my department's work. (Limited process view)	Improve the work as everyone sees it. (Broad process view)
Improvement is primarily by technology.	Simplify processes first; then use technology wherever and however it is helpful.
Maximize revenue; minimize expense.	Minimize process waste.
Maximize return on capital or RONA.	Minimize the need for capital with flexibility.
Capture economies of scale.	Capture economies of time and energy.
We do most things right.	Respond to negative feedback immediately.
React to problems when necessary.	Anticipate problems whenever possible.
Managers and staff plan; workers execute it.	Everyone participates in work planning.
Do the work to the best of your ability, but keep going.	Always think about correction or improvement. When something is wrong, stop work to think.

FIGURE 2.6
Counterintuitive concepts from the Toyota Way.

LESSONS FROM TOYOTA TODAY

People benchmark Toyota's organizational innovations, not its technical ones. For those, they are more likely to study Apple or Mercedes-Benz. Toyota is known for integrating and perfecting technology, not for flying into the technical blue yonder or distinctive styling. More typical of Toyota are less ballyhooed innovations such as an aluminum casting technology called Simple Slim, which halves the cost of an engine block.[28]

But as Toyota vehicle designs advance closer to the technical edge, the question is whether the Toyota Way will excel discovering and perfecting new vehicle technology. And there's another concern: Toyota's now-gargantuan size. Will such a people-centered system bog down in global-scope projects and operations? Can it step up to pioneering innovation? Evidence so far is indeterminate.

As Toyota expanded globally, sprigs from the TPS incubator in Aichi sprouted transplants from the home stock, testing the system in other cultures, with opportunities for mutations, good and bad. Both vehicles and production equipment became more automated and computerized. But a retiree of 30 years ago visiting any Toyota global location today would recognize the basics of TPS, although the variances in learning discipline might be disappointing to him.

Toyota became the king of commodity vehicles, not by discounting prices, but by offering value for the money. High margins let Toyota navigate through market downturns, plying a steady course of global expansion. When competitors faltered, it usually gained market share. However, global expansion stretched supply lines and bulked up inventories. Toyota's once-vaunted total inventory turns now differ little from other global vehicle manufacturers, although work-in-process remains low.[29]

But being cash-rich became a curse, a temptation to expand, which stretched Toyota's cadre of experienced senseis too thin to mentor people in its ways.[30] As early as 2000, its top leaders anticipated quality trouble; inability to hold standardized work is a precursor. Sure enough, in 2004, quality recalls increased. To get these in check, model development times were extended.[31]

Toyota is attempting to bolster the design system, the production system, and the culture worldwide. One reason for codifying the Toyota Way was to develop foreign leadership for sensei roles more quickly. Worker development is being augmented by online, faster technical education of workers, and giving them better access to best standardized work methods. These are elements of a new Just In Time (JIT) initiative.[31]

Toyota would like all production associates and technicians to acquire the equivalent of at least a two-year technical degree. Without solidly understanding current technology and software, they bog down trying to autonomously improve twenty-first-century work processes. Developing people is the limiting factor in organic growth: Not every new factory is a showcase; not all suppliers are first-rate; not all dealers are tops in service.[33] Much remains to be done.

Until the financial implosion of 2008, Toyota did well against its two most formidable competitors, Honda and BMW, both smaller and lacking Toyota's breadth of product line. The three companies stack up well in patent counts and innovation.[34] All three put quality first. And Toyota plans to make steady progress toward being an environmental leader by regarding vehicles, traffic environment, and people as the three pillars of a holistic approach to "sustainable mobility."[35] The leadership understands that this will require a shift to a new business model.

But can Toyota's tentative moves actually cope with Compression? It has grown big, let expansionary business thinking creep in, and become wedded to its old success formula in expansion. It may only have a longer fuse than Ford or General Motors. Any organization this size will have great difficulty morphing into something radically different from what it has been.

Toyota's sheer size increased pressure for financial results. In 2003, despite record earnings, Toyota stock fell. Outside stockholders insisted on higher dividends and on selling stock held in supplier companies to boost share yield—typical investor demands. The Toyota board cut its size by half, sought non-Japanese directors, and announced the Toyota Way, signed off by each member of the board. Whether many investors grasped the role of the Toyota Way in Toyota's short lead times, capital avoidance, and flexibility to operate at different economies of scale is doubtful. Most just want financial growth.

Thus, perhaps the biggest threat to Toyota is its excessive financial success from which it culturally struggled to stay immune. It won its battle to stave off bankruptcy 50 years ago.

Orthodox business analysts cite Toyota's market and financial growth as evidence of its superiority. However, in Compression, the Toyota Way, not Toyota's business success in expansion, is Toyota's greatest gift to the world. It is a seedbed of ideas for germinating a *practical approach to human work* with minimal dependence on monetary reward. But the real test of Toyota itself is still ahead. Can it conjure the spirit of Taiichi Ohno to reinvent a system to deal with the challenges of Compression?

ENDNOTES

Extended version of footnotes available at http://www.productivitypress.com/compression/footnotes.pdf.

1. Many points in this chapter were gleaned from Professor Emeritus Jinichiro Nakane, Waseda University, Tokyo.
2. The Toyota Way was seen on a visit to The Toyota Institute in 2003.
3. Jeffrey Liker, *The Toyota Way*, McGraw-Hill, New York, 2004.
4. Both the family and the company were named Toyoda. The vehicle division was named Toyota to distinguish it.
5. The term "lean manufacturing" is from *The Machine That Changed the World*, James P. Womack, Daniel T. Jones, and Daniel Roos, Scribner, New York, 1990.
6. Jim Huntzinger, "The Roots of Lean," *Target*, 18(2), 2002, pp. 9–22.
7. *The Evolution of a Manufacturing System at Toyota*, Takahiro Fujimoto, Oxford University Press, 1999, Figure 3.2.
8. Taiichi Ohno had a long-time tiff with Shigeo Shingo, a well-known industrial engineering consultant to Toyota. Both are now dead. One can only guess, but the issues probably centered on this point.
9. Robert W. Hall and Jinichiro Nakane, "Family Ties," *Target*, Fall 1990, pp. 4–11.
10. *Zero Inventories*, Robert W. Hall, McGraw-Hill Trade; ISBN: 0870944614, 1983, pp. 284–294.
11. Melissa Rumizen and Bill Baker, "Knowledge Management Based on Your Organization's Approach to Life: Customer Intimacy," *Target*, Vol. 21, No. 5, 2004, pp. 24–29.
12. Michael J. Kennedy, *Product Development for the Lean Enterprise*, Oaklea Press, Richmond, VA, 2003. Much of this section is from the work of Kennedy and the late Alan Ward at the University of Michigan.
13. David C. Kang, *Crony Capitalism*, Cambridge University Press, Cambridge, U.K., 2002.
14. Robert W. Hall, "Tokyo Sekisui," *Target*, Issue 2, 2008.
15. All societies are reluctant to admit that they have skeletons in their closets or to contradict their own folklore.
16. Figure 2.4 summarizes ideas from the writings of Edward T. Hall, *Beyond Culture*, Bantam Doubleday Dell, New York, 1976, and several of Hall's other books. Edward T. Hall is unrelated to the author.
17. Li-Jun Ji, Richard Nisbett, and Kaiping Peng, "Culture, Control, and Perception of Relationships in the Environment," *Journal of Personality and Sociology*, 78(5), 2000, pp. 943–955.
18. David Magee, *Turnaround*, Harper/Business, New York, 2003, pp. 87–89.
19. James M. Perry, *Arrogant Armies*, John Wiley & Sons, New York, 1996.
20. Observations based on 30 years of seeing "lean" companies.
21. *IndustryWeek*, a publication of Penton Media, has sponsored the Ten Best Plants Award since 1990.

22. All recent Shingo Prize winners can be found at http://www.shingoprize.org/1994 and 2001 editions.

23. Gerald J.S. Wilde, *Target Risk*, PDE Publications, Inc., Toronto; can be read at: http://psyc.queensu.ca/target

24. John Schwartz and Matthew L. Wald, "Shuttle Panel Considers Longstanding Flaws in NASA's System," *New York Times*, Apr. 13, 2003.

25. John Schwartz, "For NASA, Misjudgments Led to Latest Woes," *New York Times*, July 31, 2005.

26. H. Thomas Johnson and Robert S. Kaplan, *Relevance Lost*, Harvard Business School Press, 1987. Today an ongoing conference and forum can be tracked at http://www.leanaccountingsummit.com

27. American-based exceptions are Ventana Medical Systems and Autoliv (Swedish-owned).

28. John Lippert, "In a Supersized World, Toyota Slims," *International Herald-Tribune*, Feb. 22, 2006.

29. Richard J. Schonberger; *Best Practices in Lean Six Sigma Process Improvement*, Wiley & Sons, New York, 2007.

30. Just-Auto.com Editorial Team, "Belgium: Toyota Turns to Workers for Better Quality in Europe," Dec. 12, 2005.

31. Associated Press, "Toyota May Delay Models to Work on Quality Control," *New York Times*, Sept. 25,

32. Hirohisa Sakai and Kakuro Amasaka, "HIA, Highly Reliable Production System by Developing Intelligence Operator," paper at the *17th Annual Conference of the POMS*, Boston, MA, Apr. 30, 2006.

33. This is a legacy from old wars between dealers and the OEM's and antitrust laws.

34. "The Patent Scorecard 2003," *Technology Review*, MIT, Vol. 106, No. 4, May 2003, p. 59.

35. "Japan; Toyota Standardizes Economy Meter," Just-Auto.com, Oct. 3, 2006.

3

Learning to Learn

If we are going to deal seriously with Compression, we need working organizations vastly superior to those of today. The big potential for improvement is in learning to do much more with much less. This chapter will explore how to "learn how to learn," in three parts:

- Individual learning
- Process learning—revolution of man-made processes or influencing natural ones to do much more with much less
- Organizational learning by more effective systems and behaviors

Just learning that Compression is upon us is no small intellectual challenge—and an emotional one. Our instincts from the expansionary era scream that it must really not be so. Once convinced that Compression is real, there is no pat, once-and-for-all answer. We must personally think through everything we do—and rethink it—for none of us alone can begin to comprehend the details of complex interrelationships in our technological society. For this, we need fast-learning organizations, able to dive quickly and deeply into technical, natural, and man-made processes (or systems) and connect the dots. We have to become better not only at analysis, breaking things down, but also at synthesis, integrating them together, moving quickly beyond amazement by drawing big diagrams exploring newly seen systems connections from a wider world.[1]

We associate learning with school, but our workplaces and all the rest of the world is a school, open 24/7. Many schools offer a Learning to Learn minicourse. The first tip is to keep a regular study time every day, which few students do because it conflicts with other things they prefer to do. They are learning, but things that they like in ways they like. For instance, actively learning a video game based on medieval mythology is more fun than passively learning medieval history from a lifeless book. Learning can be a tough discipline.

Both in sports and in music, cumulative practice time over one's total life highly correlates with performance. Talent, acclaim, and big bucks help make long hours of practice less onerous, but many of us also squander time and money on a hobby we truly enjoy with no prospect of either acclaim or payback. That is, some learning is attractive while other learning is more like work. It helps if one is intrinsically interested.

Economic systems parallel dominant thinking patterns and therefore learning patterns. Industrial society fragments everything into separate pieces of property that owners can trade or manage. If private property is a fuzzy concept, people have to think more about relationships and interactions. We need more balance in our thinking.

IMPROVING YOUR LEARNING PROCESSES

Learning processes can improve by two routes: (1) speed—faster feedback and correction of errors, and faster detection of precursors of change (positive and negative), and (2) expanded scope—discovering phenomena not known before and associating previously disconnected concepts (which is one definition of creativity).

These routes are much like Chris Argyris' double-loop learning, which he applied mostly to strategic planning by executives.[2] One can extend his idea to four loop levels:

Correction: Sense and correct errors with little or no change in process.

Prevention: Identify and correct the causes of error (precursors) to prevent recurrence; this is often called fail-safe or *poka-yoke* (Japanese term) in operating processes. The same lessons need not be learned over and over.

Anticipation: Identify precursors of change earlier and in a broader context.

Expansion of context: Search outside the system of interest to project changes from past patterns; seek how to improve the learning processes themselves.

In learning physical skills, hunters learn to shoot ahead of the duck, and athletes may develop an uncanny sixth sense to anticipate where both ball and players will go. But in business, forecasting item sales by extrapolating prior sales data works well only when buying behavior is stable. Quantitative projections cannot extract more information from a data set than is intrinsically in it. Smart companies do market research outside their prior sales history. In Compression that is still not enough. To serve all stakeholders well, including the global environment, crucial work organizations need to consider much more, learning much more, much better and much faster.

PART 1: INDIVIDUAL LEARNING

Although what we need is better collective learning, human organizations are a collection of individuals. Understanding how to improve learning begins with better insight into the strengths and weaknesses of the basic building block of organized learning: the individual.

Neural Learning Mechanisms

Gregory Bateson once posed a great question: "What is a man that he may know anything?"[3] Physically, we are bodily functions and sensory organs, most of which connect to a brain through neural pathways. Brains learn by comparing present sensory experiences with a stored history of similar experiences. These evolving neural patterns are sometimes called engrams.

The brain sifts new experiences to decide whether using them to adjust already-established engrams will help predict something deemed important. Adult learning consists of modifying old engrams to greater or lesser degrees. We revise engrams to improve predictability, so when a new observation indicates that one has been totally wrong, it devastates our confidence and we try to rationalize it away.

A prediction can be as simple as anticipating the stub of a toe, or as complex as Einstein's insight that $E = mc^2$. In any case, engrams imprint a chain of causal logic by which to interpret new information. The brain constantly assesses incoming information to see whether the cause-and-effect logic in engrams should be modified—or maybe even erased and started over. Effective learning lets us make better responses when a similar pattern is seen again. Crucial to betterment of individual learning is improving the logic for revising engrams, including positive and negative emotions attached to prior changes.

Engram memory is associative. Connections between the right places must occur at the right time. The chemistry and energy flow in the brain regulating this has been called Hebbian learning: repeated firing of one neuron triggers others nearby.[4] Greater sensory stimulation excites more neurons and juices metabolic energy to connective nerves to power still more neurons to fire. En masse, neurons firing in waves generate electrical energy, lighting up brain scans showing where these waves are concentrated. A strong stimulus, like a loud noise, or from multiple senses at once (like sight, sound, and smell) excite more initial firing, and thus bigger waves of activity. Big waves make enduring changes in engrams.

Intense experiences are seared into memory by neural wave tsunamis while something boring doesn't make a splash. That is, there is a physiological basis for an old tongue-in-cheek, four-stage description of human adult learning:[5]

1. This is worthless nonsense.
2. Interesting point of view, but perverse.
3. May be true, but quite unimportant.
4. I always said so.

Any ideas to improve adult learning must contend with sluggishly shifting engrams, including the emotional ones. They must also contend with limited and sometimes misleading sensory capacity. Human brains integrate our five basic senses into a combined pattern that makes sense, none of which are anything special. Bats use ultrasonic frequencies to fly in the dark; elephants communicate (and sense distant storms) using ultra-low-frequency sounds; and something warned most animals to flee the Asian tsunami in December 2004, while humans oblivious to the receding sea drowned.[6]

Human eyes are extensions of our brains, but they detect only visible light, only 5 percent of the total electromagnetic spectrum. Approximately 260 million rods and cones in our retinas detect photon patterns, but only 2 million ganglion cells funnel them to the brain.[7] Because the brain processes only a fraction of what the eyes detect, it must tell the eyes what to concentrate on, so part of learning concerns what to pay attention to.

Multitasking overloads neural highways, and batting a 90 mph baseball demands synaptic speeds that very few people can cultivate to the major league level. In a technical world, much of what we see is detected by extrasensory technology converted into a form our ordinary senses can pick up. These are easily misled even when observation has been scientifically designed.

For example, the astronomer Percival Lowell for decades insisted that he saw "spokes" on Venus, much like the canals on Mars sketched by earlier astronomers. However, the 1972 *Mariner 9* flyby of Venus showed no sign of spokes. Conjectures abounded. How could Lowell have been so misled? In 2002, a few optometrists who happened to be amateur astronomers noted that Lowell had narrowed the aperture of his telescope to three inches or less, an unusual setup that mimicked an ophthalmoscope. Lowell most likely saw a reflection of the blood vessels in the retina of his own eye.[8]

The well-known video exercise of a gorilla crossing a baseball field illustrates how our concentration can draw us off. Viewers who are asked to concentrate on counting exchanges of the ball almost never spot the gorilla.[9] Deceptive technology is now so easy that we routinely question whether digital photos have been Photoshopped. On the other hand, lip-synching on stage is so normal that an unfiltered voice seems unreal.[10]

We detect sensory misperception if clues don't jibe. For example, most people self-infer that a vivid experience of being abducted by aliens was actually a dream.[11] The experience is out of kilter with stored engrams, so we must have dreamed it—but we use this same mechanism to predict the future, to dream dreams of what could be. Unfortunately, this same ability also fools us into seeing things that don't exist and vice versa. For example, visual illusions work by calling the wrong engrams from our neural files, so distinguishing fact from fiction is not as straightforward as it might seem.

Much of our learning in an industrial society has to be derived or inferred from symbolic media. Not only are these media subject to distortion, but so are the neural mechanisms by which we interpret them. For fast, high-quality learning we need rigorous standards for accepting new observations as fact, not whatever we might wish to be true.

"Natural" Learning

Some adults regularly confound imagination, dreams, and reality. The Yanomamö of the Amazon were an example. Sixty years ago, they had no means of hyperrational learning: no books, no schooling, no vehicles, and no electronics. Now and then a shaman's conjuring of powerful spirits worked up one village to raid another, usually killing all males of any age. Females they saved to expand their gene pool—and their labor pool.

Yanomamö narratives abruptly changed channels, mixing events sensed in their spirit world with physical events actually witnessed, and possibly juiced a bit by snorting ebene, a psychedelic. In such a violent environment Yanomamö increased their chances of survival by learning sixth senses. Mark Ritchie marveled at how they read devious intent in others by the "spirits living in them." For example, an ex-shaman avowed that a teacher in a missionary school was a molester long before this was discovered by the missionaries.[12] In dangerous street situations, people in industrial societies must call on the same sixth sense. Not everything we learn is by scientific experiment.

Yanomamö engrams must have formed by blending revelation from their spirit world with real experience. But how different was their learning process from that in much of the industrial world? Could hyperrational media serve a learning role similar to the Yanomamö's spirit world? For those who assign credibility to anything seen on TV or the Internet, perhaps it does.

Yanomamö learning evolved in the little tribal world where they dwelt. They had no way to see themselves in any larger context until outsiders opened it to them. (Whether this truly benefited the Yanomamö has been debated.) Today, many in an industrial society are similarly stuck in a commercial transaction world, but being in it, can't see it in any larger context. Axiomatic assumptions of business and economics aren't that different from tribal lore. To progress, we must question that tribal lore. Openness to discovery is essential for learning, but so is a methodology to assure that learning is real. If work organizations consist of tribes at war, like the Yanomamö, learning doesn't progress much beyond how to survive in that locality.

Proof of learning by the scientific method has to be explicit and logical, but as with the Yanomamö, learning is not strictly scientific. Some of it is tacit, like a surgeon learning exactly where to insert a pacemaker probe at the right moment. Indeed, learning that we can't relate to direct experience (like abstract math) bores most of us, but stories with emotion motivate us. The motivation to cope with Compression is mutual survival; but as long as its challenges seem abstract and far removed, they are not as emotional as the immediacy of financial challenges that we set for ourselves.

Even in scientific reasoning, emotion plays a role. Without persistently seeking causality, it is easy to attribute all problems to evil spirits or other people. On top of that, we're easily fooled into thinking that hyperrational media and rational models are infallible.

THE LIMITS OF RATIONALITY

Except mathematicians and philosophers, few people know of Gödel's theorem. In 1931 Kurt Gödel drove the final nail in the coffin of rational determinism, the belief that rigorous logic or computer models can solve all mystery. At the time, logicians still hoped to create an axiomatic mathematical framework that could logically explain everything.[13] Gödel's theorem took the air out of the sails of this quest, proving by mathematicians' own logic that extending our limited perception by axiomatic logic to frame all knowledge is impossible. Others later proved the same thing.[14]

Gödel's theorem does *not* prove that we can never know anything. It proves that no logical construct can represent *everything*. Any hyperrational model merely represents reality. It is not the reality it represents. Some models are just for entertainment; others attempt to explain reality to human minds. Many of these models are essential to us, but none is perfect or complete. After deflating rational determinism, Gödel's theorem became mostly a backdrop for paradoxical logic exercises.[15]

In brief, Gödel's theorem goes like this: To any system of formal arithmetic a proposition can be put that cannot be formally proved or disproved within the system. Why? The system is either internally inconsistent or it is incomplete. Gödel's arithmetic used metanotation, representing logical constructs as integer numbers. Therefore the proof covers all logical constructs and all models.[16]

Accounting, finance, and other models by which businesses are run are all subject to Gödel's theorem. Any quest for a perfect system is hopeless. Elegant sensors, rigorous quantification, or cool simulations cannot fully represent infinite reality. This limitation is rarely mentioned in schools of business—or schools of engineering. However, the human mind is a sucker for the illusion that a model or a system is the reality it merely represents.[17] No model can answer questions about itself or outside itself. Attempting to do so is sometimes called self-referencing.

Gödel's point has been made less rigorously for centuries using paradoxes from Zeno the Greek onward. Zeno's race between the tortoise and the hare was a paradox to illustrate the futility of reaching infinity: A hare going only halfway at a time takes forever to travel any distance, so once behind, a hare can never catch a tortoise.[18] Oriental Koans hint that reality is beyond grasp: "The more you pursue (study by the mind), the more does it slip away."[19] Several of Christ's parables illustrate absurd contradictions of absolute law; so the flexibility of love comes nearer to covering infinite complexity. That is, no system of logic is capable of perfect prediction, but idealists craving magic answers still look for loopholes.[20]

Self-referencing is accepting the validity of a source that only references itself. For example, should we trust an odometer reading that a ten-year-old car has only 30,000 miles? Experience with other vehicles tells us to be skeptical. The opposite of this is absolutism, unshakable faith in nonvalidated logic that a belief (model) will always work for us. A tribal shaman invoking a more powerful spirit is obviously self-referencing, but is an entrepreneur seeking to make a dream become reality much different? We can only hope to train our brain to make us self-aware when we are self-referencing.[21]

The limitations of explicit models, including those of language, are important for all learning: individual, organizational, and process. Personally learning the systemic interconnections of all processes in a technical world would take longer than any human can live. On the other hand, if we learn only fragmented, technical chunks, we can't see how they relate to anything else. Even worse, superficial sound-bite familiarization lets us show off vocabulary without really understanding what it represents.

For fast learning we have to beware of false rationality. A great deal of what we need to know, even in highly technical areas, is implicitly absorbed; or as Yogi Berra would say, "You can learn a lot by just looking." But to do that, one must spend time learning how to look. Offhand, that sounds stupid, but learning what to look for is not always simple, as with doctors learning to diagnose patients. For instance, much damage can be done by people who know how to program, but foggily understand the processes that they are trying to augment with software. Contextual learning is very important. That's why Toyota maintains the maxim of going to the physical scene of a problem whenever possible.

We deepen our understanding of reality if we constantly seek it while avoiding the traps we set for ourselves in the models we use to describe it.

Extreme versions of the deconstructionist philosophy hold that reality does not exist, that it is all some individual's paradox-laden neural imagery; therefore, if no person's explanation of it is any better than any other's, no proof of anything has validity.[22] However, these doubters don't test the existence of reality by stepping in front of a speeding beer truck, and we have no choice but to use models. If we are wary of both our weaknesses and our models' flaws, our better models do convey useful information.

Integrative Learning

Twenty-five years ago, Howard Gardner proposed seven different intelligences: linguistic, logical-mathematical, spatial, musical, bodily-kinesthetic, interpersonal, and intrapersonal.[23] Classroom education generally appeals to linguistic and mathematical skills, but learning may involve any or all of them. Most of us are stronger in one or two intelligences in this mix and prefer activities in which we can do well. Most experienced classroom teachers can relate to a roomful of children having diverse intelligences.

Because we live in a hyperrational world, individuals in learning organizations may need a modicum of linguistic and logical-mathematical intelligence to participate as a professional in innovation and improvement. They will need all the other intelligences also, but the intelligence often overlooked is intrapersonal intelligence, or self-awareness, the ability to see our own effect in a work environment. This may be hard for some folks, and not all of us are cut out for it, but many advances that we call civilization came through a long struggle to subjugate our natural instincts.

Coping with Compression requires many people in any work environment to see a bigger picture of the world than in the past—to understand how both processes and systems work from multiple viewpoints, not from a severely truncated overview, as when trying to understand General Motors from a simple set of financial statements. That's integrative learning. It goes by many synonyms: systems thinking, process thinking, holistic thinking, and the like. Western interest in this dates at least to the 1950s, when Forrester, Beer, Ackoff, and others began intellectual study of how organizations perform as dynamic, interactive systems.[24]

A modern version of this is system-of-systems engineering, which recognizes that a complex system that includes humans isn't a totally explicit structure. This approach seeks adaptive models, recognizing that people, computers, materials, space—everything—melds in an overall system.

However, given the rate at which engrams can shift, whether we can "brain" our way into change faster than Toyota can "create" it through the Toyota Way, is problematic.

Our daily media deluge illustrates another learning challenge: integrating disconnected fragments screaming for attention. In some contexts, individual skill doing this can be acquired. For example, chess grandmasters instantly read chess formations in chunks, comparing them with engrams of previously seen patterns of moves and countermoves. The rest of us see a mass of detail, trying to imagine a move or two ahead from every angle. Physicians adept at diagnosis also appear to see patterns of symptoms as chunks. Acquiring such ability is thought to take ten years or so of intense absorption in a field.[25]

Part of human development has to be learning to connect dots in a greater context, cultivating discipline to regularly stop and reflect—to pull our engrams together. For example, will volume production of a bromine compound help or hinder phytoplankton photosynthesis?

Absolutism and Denial

Medieval Catholic Church inquisitions tested all truth against prelates' interpretation of either scripture or ancient wisdom. These absolutists' comforting faith that ancient doctrine could not possibly be wrong blocked acceptance of new discoveries based on evidence. Independent thinkers running afoul of these tests for absolute truth included Copernicus and Galileo.

Consequently, today those slow to revise neural engrams are apt to be derisively called a "flat earth society" or "in denial." This condition is a by-product of the way the brain learns, so all adults exhibit some degree of "denial lag."

All humans' learning cycle corresponds to their life cycle. Having seen it all, adults modify old brain patterns more than they create all-new engrams. They do appear to better integrate previously learned patterns—that is, acquire wisdom—but if they cannot dispel using old patterns to explain new observations, they miss much that is in plain sight.[26] Adults creating totally new neural patterns need to grow a different set of axon associations. In an extreme case, adults blind from birth who suddenly acquire sight endure literal pain as their brains try to reconstruct.[27] Learning a foreign language is tough at an advanced age.[28]

Each human can only learn during *one's own individual* life span. Each infant starts anew, easily deceived that the entire world resembles what they see around them, and maybe always did. For example, in 2007 authorities feared the rising use of crack cocaine because youth unaware of its scourges twenty years earlier figured that it can't be worse than meth, the scourge of the present.[29] Using similar logic, every generation repeats many mistakes of its elders, as is well-known by parents of teenagers. Some of history's most useful lessons, if known at all, don't stir listless neurons while the heroics of relearning them the hard way create a neural buzz.

Regardless of age, we're all subject to the illusion that our prior experience represents all reality. A dramatic exception was Traudl Junge, cloistered for four years at age twenty-one as Hitler's secretary in his bunker. Few Nazi insiders ever escaped their own self-denial, but once out, Junge saw the world anew, regretting her blindness to anything outside the unreal little world in which she had existed.[30]

Susan Choi's case is more prosaic. As a youth she dwelt exclusively, 24/7, in artificial, air-conditioned, teenage hangouts in Houston. Only after leaving could she see that these commercial cocoons weren't the beginning and end of all reality. Houston had parks, but Susan did not see them.[31] This in-the-box effect allows Westerners in industrial societies to see their high-consumption world as normal, never questioning how other people might even frame it as terrorism.[32]

PART 2: COMPRESSION OF PROCESS LEARNING

Process is an all-encompassing, analytical word, defined as a combination of smaller steps by which anything is done. Bill paying is a process. Beer making is a process. Photosynthesis is a process. So is ocean acidification or formation of a galaxy. Learning is a process, too.

A process is as big or small as you conceptualize it to be—we can never unravel all its connections to a wider universe. Flow diagrams and other hyperrational aids help humans conceptualize processes that they can't sense directly all at once.

Some processes, like bill paying, are almost totally human. Some, like fermenting beer, are natural processes controlled by humans. No puny human is apt to affect the enormous natural forces in the process of forming a galaxy; but on earth, all human work can be described as a process, and many human processes interact with natural ones.

So, what is process learning?

- Humans improving their understanding of how a process works.
- Making an existing process more effective for human purposes by better control or by eliminating waste using a broader definition of waste than discussed in Chapter 2.
- Creating new processes to benefit humans. New processes may or may not displace existing ones.
- Improving human intervention to shape natural process evolution in a direction more favorable to humans, as they then perceive it, so learning what is "good for us" is an important part of this learning.

Almost any action to understand or improve a process can be thought of as problems to be solved. Thus problem solving and process learning are intertwined.

Many of our most common process phenomena are not well understood, so there is no end of things to learn. An example is the basic chemical and physical properties of water. The unusual physical properties of water have promise so far untapped, and almost all biology is water-based chemistry in which the role of the water medium is not well understood.[33]

Improvement of an existing process by small-step kaizen is now well understood: first eliminate unnecessary steps; then improve execution of some of the steps themselves. Changes may be procedural, technical, or a combination of the two, and rounds of improvement are never done. Better is always possible, so set up work processes to be learning labs: high-context, visible processes that let observant people see clues to further improve them.

But this presumes that we understand what "better" is—what is truly good for us—but that knowledge is never complete either. Agreeing on what is good for us is rarely by consensus, so learning must also seek better ways to work out wicked problems.

A Broader Definition of Waste

The list defining waste (as seen by lean operations practitioners) in Chapter 2 is insufficient for Compression. We need to add to that list by considering how nature sees waste in order to reach well beyond waste as only a customer might see it.

Errors, defects, scrap, and rework: Eliminate all errors, down to zero if possible.

Nonunderstanding: Clarify the purpose of a product or process to avoid the waste of complexity in transactions, records, designs, or unnecessary process steps. Also called "imaginary waste," which is avoidable if we more clearly understand what we do and why.

Virgin material waste: Use no more virgin raw material than necessary. Reuse, refurbish, remanufacture, and recycle before using virgin material.

Energy: Use no more energy than necessary. Draw no more energy from a source outside the immediate environment than necessary. Find uses for waste energy streams. Try not to use energy just to warm nature (waste heat). Strive for a high overall energy yield.

Remediation (of natural states): Anticipate and prevent waste at the earliest stage rather than create more waste by remediating messes later. Avoid releases of toxic substances that must be cleaned up later. Use no more material and energy than necessary for a vital purpose. Thinking that environmental cleanup can only be through remediation appears to be a major reason why business managers think that environmental sustainability will drive them bankrupt.

Waiting (long lead times): Any material, idle equipment, or person waiting for something to be done. Energy is being consumed doing nothing.

Processes themselves: Don't waste resources on operations that need not be done at all, and eliminate the imaginary waste from those that are necessary.

Motion: Excess motion adds no value (except perhaps in show business). Minimize the motion necessary, and reduce the distance of that which does seem necessary.

Transportation: Related to motion; unnecessary movement by road, rail, air, sea, or donkey wastes fuel, vehicles, roads, and time. Do more in the same locations.

Inventory: Inventory is usually a monument to other wastes that allow it to accumulate. The time, space, and energy necessary to manage inventory itself are wastes. Exceptions are inventories to cope with seasonal changes, and compact inventories of prior knowledge that catalyze further elimination of waste. In nature, seed banks are examples of compact knowledge inventories.

Performing when not needed: Overproduction in manufacturing is an example, often done because the system begs us to. More generally, do the right thing when it is timely to do it. When working just to be learning, try to minimize resources consumed. (Pilot training with flight simulators is an example.)

Information: Keep it correct, simple, useful, and accessible where needed when needed.

Human nondevelopment: Don't waste any person's potential if it can be helped.

This list extends the one in Chapter 2 to meet the challenges of Compression, just as a guide to thinking, for the wastes overlap and have exceptions. In general one can describe elimination of process waste as simplifying a process to its lowest overall energy state.

Big-Step Innovative Learning

Big-step process learning invents processes never seen before—called revolutions or innovations if they change how we work or live. Historians laud many product examples: steam engine, telegraph, photography, telephone, autos, radio, TV, photocopier, personal computer. These objects came to symbolize whole new industries with processes to develop them, make them, distribute them, service them, use them—and dispose of them at the end of use. But all these processes were seldom in the same company, which obfuscates understanding of a full life cycle of a product or service.

The new processes created new sets of problems, many of which were not foreseen when they began, and many were done by fragmented work organizations competing in a market system. Today, a big-step innovation still creates fledgling processes that need development by small-step improvement. A different business model is often needed for them if the new processes have to attract more income from customers than is spent on processes to serve them. Changing a legacy business model is traumatic for a work organization encrusted in the engrams of people set in their ways, which makes this the hardest part of its innovative learning.[34]

Many small-step process improvements are trivial, like moving the trash can closer to the mailbox. But such changes should provoke deeper questions: Why so much junk in the mailbox? Probing that leads to questioning larger business processes. Of course, one recipient can't stop a horde of mail houses from dispatching their promotions, but a protest movement of recipients may induce the advertiser clients of mail houses to find a less wasteful way to generate prospects.

The cumulative effect of many small improvements may become the equivalent of a major innovative change. Other times a big-step innovation must displace a well-improved old one that has no more potential. An example is converting a paint system from solvent based to water based. Unending improvements in a solvent-based system will never lead to a water-based system.

Process learning covers a wide range of change. Process learning also hints that it depends on the personal and organizational development of the people improving the processes.

The Implications of Learning Curves

The curves in Figure 3.1a go by different names: learning curves, experience curves, and progress functions. In popular use, people often refer to a learning curve as if it were rising, but the curves actually used slope downward as shown.

Process learning curves (see Figure 3.1a) graph or project process improvement by some criterion, often workdays per unit finished, so smaller is better. Sometimes the curves are called *progress functions*. The

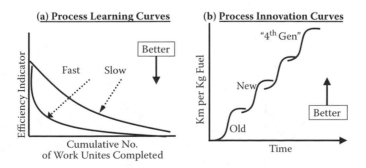

(a) Process Learning Curves   **(b) Process Innovation Curves**

FIGURE 3.1
Graphic models of process improvement.

horizontal axis is not time, but the cumulative number of work units completed. Improvement decreases by some constant percentage each time the cumulative number of completions doubles, so improvement on each unit is much greater early on than later. Learning curves are old hat to readers in businesses such as defense contracting, built into contracts and bid estimates. Others may never have heard of them.

By contrast, process innovation curves (see Figure 3.1b) are Gompertz curve depictions of innovative changes over time. In this case the y-axis is a design criterion for the product being developed, like fuel economy or the thrust-to-weight ratio of aircraft engines.

A plot like that in Figure 3.1a will show some improvement without any special effort to improve anything. Plotting completion times of a person beginning to learn a skilled manual task will produce a similar curve called an experience curve. Aggressive process improvement should produce a much steeper curve than just letting it happen.

Learning curves of this type originated in 1936.[35] They take the form $y = ax^k$, as in Figure 3.1a, but in logarithmic form they are $\log(y) = \log(a) + k \log(x)$, which graphs linearly as shown in Figure 3.2. Learning curves map small-step improvements.

The equations for a learning curve follow a general relationship called a power law. A power law fits data from many other phenomena as well, implying a relationship among them. Possibly the best known power law offshoot is the 80/20 rule, or Pareto distribution (e.g., 20 percent of a population gets 80 percent of the total income, with everyone getting some).

However, similar power law functions have also been found to fit many other kinds of data and ratios in numerous settings, notably:

- Scaling of connectivity links in a self-similar network: $P(k) = Ak^{-y}$
- Boltzmann entropy: $I = K \log W$

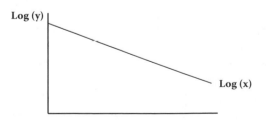

FIGURE 3.2
Log–log graph of a power law relationship.

- Shannon information entropy: $I = -\Sigma\, p_i \log p_i$
- Kleiber's law for biological scaling: $B = B_0 M_b{}^b$ (e.g., metabolism to body mass)
- Zipf's law of word use frequency in languages: $f = 1 \div [k^s\, H_{N,s}]$

Plots of data for each of these phenomena on a log-log scale all look like Figure 3.2. In recent years enthusiasm to see power law relationships in all kinds of phenomena may be overdone, but characterization by a power law does suggest that each of these phenomena shares the characteristic of having patterns of partial self-similarity at all scales. Note that the examples connect thermodynamics, information, biology, and language.

Geometric figures that exhibit a self-similar pattern at all scales from minute to enormous are said to be self-scaling. Examples from nature are fern leaves and snowflakes, which are also artificially generated by fractal equations as displays on computer screens. The classic computer-generated, partial self-scaling geometric form is the Mandelbrot set, which can now be seen at dozens of Web sites.

So why is this of interest? Partially self-similar patterns suggest intelligent communication, perhaps even learning, as illustrated by Zipf's law for natural languages.[36] Word frequency counts fall into a power law distribution because language—including acronym-laden jargon—conveys meaning by connecting unexpected symbolic concepts using common syntax patterns. Brains receiving a message constantly compare an incoming pattern with what they expected. When a symbolic interjection differs from that pattern, the intelligent processor has to decide how it fits with whatever else is stored in memory.[37]

Neither biological nor artificial brains can communicate without sharing a common syntax and a common set of word meanings, or vocabulary. Because engrams fill in the gaps, transmitter and receiver need not share a precisely identical set. A robust syntax helps establish a common context for messages sent between them. Human languages differ in syntax and grammar as well as vocabulary. Evidence that humans create different syntaxes using universal rules is strong, but not yet conclusive.[38]

If every word were identical, language would be a meaningless pattern, like a metronome ticking. If no two words were the same, we could not interpret anything either. Likewise, if all words were used with equal frequency, very little meaning could be imparted. Only patterns of regularity connecting irregularity convey heavy-duty meaning. Researchers hoping to detect intelligent life in deep space analyze space noise for patterns that follow Zipf's law.

For example, a plain brick wall, every brick identical to every other, is boring; nothing sticks out. Graphic designers try to vary patterns so that somebody actually notices, and maybe even remembers. An artist that uses brick as a medium would try to organize it into a pattern that conveys meaning or evokes emotion.

Compression and Increasing Complexity

Were there no resource limitations or environmental constraints, the advance of complexity alone would require new concepts for organizing work. The challenges of Compression make many of these problems even more complex, so that just juggling them all in mind taxes any of us. We have to learn how to resolve major problems, both broadly and in detail, when none of us can know every detail and without letting our emotions destroy us in divisiveness.

We even need a different definition of efficiency: minimal energy processes with high free-energy density. Simplifying complexity moves processes in that direction. Explanation of this begins with astrophysicist Eric Chaisson, who among others observed that the evolution of everything seems to proceed toward increasing complexity.[39] This is supported by some evidence, and to explain it, Chaisson proposes the idea of free energy density, which begins with free energy, that which a system can apply outside itself. If the system is an animal body, free energy is used for work or play above basal metabolism, the energy needed to keep itself alive.

Free energy has an analogy in human work organizations. An organization's basal metabolism is energy used for internal maintenance and communication. All energy above that can be applied to work serving society—or wasted—but an organization that spends all its energy on internal administration (or politics) is thermodynamically bankrupt. If it were an organism, it would die.

Energy density is total energy flow, per unit of mass, per unit of time (ergs/gram/second). Free energy density is free energy flow per unit of mass, per unit of time. In nature, the highest known energy density is in the human brain. There, zillions of minute neural energy shifts create much higher energy density than in the sun. The sun is much hotter than any brain, but solar heat is comparatively simple radiation, whereas the brain's tiny free-energy gradients create abstract, symbolic order from disorder. It creates much more information using a lot less energy than the sun. In technology, microchips now have higher free energy than the brain. However, the human work processes we integrate using them may have low free-energy density—maybe even negative free energy.

Biologists and ecologists propose similar thermodynamic concepts to tentatively explain both evolution and ecological imbalances. Biologist Stuart Kauffman proposes a similar energy density concept: [amount of work done] × [diversity of work done], noting that both organisms and ecological systems appear to maximize such thermodynamic measures when evolving to stable state (balance or maturity). Kauffman and others theorize that life can spontaneously arise from collective autocatalysis in a chemical soup of diverse molecules. Of a near-infinite number of molecular combinations, only a very small number can collectively generate enough free energy to come to life. Once this happens, the nature of the system that self-organizes is not predictable; it takes on a life of its own. Kauffman thinks this not only drives another nail in Gödel's coffin for rational determinism, but also opens a bridge to reconcile science with spiritualism.[40] It also has a message for organizing work.

For our own survival dealing with complexity, human work organizations have to evolve toward becoming collectively autocatalytic. In a limited way, organizations that regularly and spontaneously compress complexity by process kaizen increase their free energy by people engaging in autocatalysis—one idea sparks a synaptic cascade in others (see Chapter 2). To do this we have to start thinking of work organizations as learning organizations, not as money machines, institutional monuments, or fiefdoms run almost like slave plantations.

Energy, Information, and Process Learning

Factory work should not be hard to see, so differentiating processes by degree of self-similarity is easier. A production line making a single identical product, like the proverbial Ford Model T line, is perfectly self-similar. However, repetitive tasks at every workstation, perfect for robots, are brain deadening for humans, an effect sometimes parodied in entertainment, notably in Charlie Chaplin's movie *Modern Times*.

Introducing product variations into this line flow makes the self-similarity less monotonous. Mix a few custom-engineered options into this sequence, and self-similarity must encompass more complexity. Vary both lot sizes and production rate and self-similarity becomes foggier, like a basic melody from which jazz players riff into their own zone. This progression is like a boring Gregorian chant transitioning into frenetic improvisation until a common theme no longer ties it together. Indeed, when building complex, unique products, less time may be spent doing physical work than in understanding exactly what must be done and why. Think of simplifying complexity as trying to make a variety of adaptive outcomes come from a simple thematic process.

Improving physical work processes is good training if they are not highly technical and easy to see. Unfortunately waste seen there often points to bigger wastes in learning and communication leading up to the physical work. Those wastes are harder for participants to detect if they are loaded with human emotion and game playing.

A way to simplify this is to regard a total process, including its information and learning systems, as a living syntax system, or self-similar generating function. Companies with well-done core designs for product families do this today. For example, online design software allows customers to customize dimensions and other variables, like materials, and check whether their design will work in their application, but prevents them from designing something the system can't build.[41]

Imagine doing something similar for a designer pharmaceutical molecule intended to attack a particular disease for a specific individual. Ingenuous learning is to discover a great deal of useful complexity, then reduce it to a simple process for execution. Only a well-developed learning enterprise can begin to tackle such challenges.

Mastery of the mass of detail necessary to do anything like that is beyond the comprehension of any one lone genius. Such processes may be as automated as an Intel semiconductor foundry, but humans still have to diagnose and fix them when they can't fix themselves. Most important, humans have to devise such systems and improve them to better serve human purposes.

In a well-integrated system, all subprocesses reinforce each other, as in biological systems, so that many people have to integrate their detail into the whole. Programmers on big projects have to do this just to make software work. To make it fulfill its purpose, they also need to understand the applications environment that the program affects, long-term as well as short-term. And they would prefer to create a platform that is easy to update and adapt, not a one-shot dead end. To do this, both software programs and work systems have to be regarded as living systems, not once-and-for-all solutions.

Think of designing work processes to be part of a learning language, so that anyone seeing a process in context knows what to do, as when using visibility signals in lean operations. But beyond this, we need prompts to see what isn't working optimally when in use. Vehicles, for example, could be retrofitted to optimize combustion on each cylinder over a lifetime, detecting when not running at best fuel economy or spewing nitrogen oxide—tough, but not technically impossible, and only a stopgap before coming up with better, simpler environmental solutions.

Process improvement typically simplifies a process low in self-similarity to one that is higher. If starting from scratch, try to devise a simple core system that minimizes complexity by devising self-similar variants on the basic pattern.

A Tree Doesn't Grow to the Moon

No tree grows taller than about 400 feet because the forces that carry water to the top reach their limit. These growth-limiting forces usually follow power laws like Kleiber's law. For example, if a 200-pound, 6-foot man grew to 12 feet but kept the same proportions, he would weigh 1,600 pounds and his bones would break trying to carry his own weight. In every living thing, the energy generated to produce growth eventually bumps against the constraints of physics, and is thought to usually reach a size close to that which lets the organism function with a minimal use of energy.

Simplistic business thinking says that bigger economy of scale is better. By spreading overhead across more units made, shipped, sold, or serviced, cost per unit should decrease and profit per unit increase. That is quantity-over-quality thinking.

But experienced managers recognize that scale-up of a system usually requires changes in processes and control systems. As small, informal entrepreneurships grow, they take on more rules and formal hierarchy. A single operating unit transitions into multidivisions, then into a conglomerate of different business models. As it grows, it gains mass and loses flexibility. Internal communication across rigid boundaries has lower and lower signal-to-noise ratios. Or as recognized by at least one business writer, excessive growth fuzzes the self-similarity patterns of organization that were effective at a smaller size.[42]

Quality thinking in developing an organization stresses how it can do what is needed very, very well, and learn to do it even better. Not only the number of people, but how they are developed and the systems they use, affect that. And a complete work organization is an enterprise that includes suppliers, customers, and others that must communicate at least on occasion. In business today, this kind of thinking is implied by right-sizing for peak performance, not maximum revenue or profit.

Another power law listed earlier is Shannon entropy, often restated as a signal-to-noise ratio, which is easier to relate to. In electronic communication this ratio can be increased by taking out redundancies like blank spaces and repeated strings, using standard abbreviations, and so on. This approach is used to compress zip files in e-mail attachments and in much other digitally encoded information.

Humans cannot easily interpret highly compressed information. However, people who text message often compress their communication with symbols. For the same reason, acronyms multiplied during the twentieth century and all work organizations and professions use them in their jargon: to communicate more, faster. Without learning these symbols and their context, an outsider is lost. Compression of information for humans is also done by making key information easy to identify in a picture, map, scene, electronic screen—or by the visibility layout of a workplace.

The size of an effective communications network is not just its number of potential connections. A phone can potentially connect to any other phone. Instead, effectiveness is the number of human nodes that can actually exchange useful information, which depends on many factors, but network size is far from infinite. Many of us are swamped with e-mail noise—too much time sorting out what we want to know. Our own input/output rates limit us even when we use well-defined search patterns for sorting and aren't just browsing.

Information and communication consume time, material, and energy. Trees are cut and electricity is used. However, the biggest waste is from delays, confusion, and corrective action when information does not result in coordinated action. Better control systems for the power grid or vehicular traffic can save a lot of energy.

Human communication is integral to efficient learning. It is a factor in every control or communications network. Shannon entropy (information) and Boltzmann entropy (thermodynamics) are both characterized by power functions, and both are akin to the general equation characterizing scaling connectivity in self-similar networks. These relationships suggest that even human communication and learning are thermodynamic processes. And by the concept of free energy density, we can do much more using much less by increasing the free energy density of all kinds of processes, from human learning to mechanical work.

Self-similarity characterizes self-organization, and at some point self-learning. A self-learning system has to become capable of second- or third-loop learning to figure out how to better self-organize. And self-organization seeks to more effectively use collective human energy and the energy used in the processes it controls.

Forget equations. Just watch birds feeding. Their objective is to use the least energy to obtain the most calories. Smart birds find the richest food sources first. When these deplete, they look elsewhere (like humans hunting oil). If food sources are short, they conserve energy or migrate to a better source. They can only store so much fat, and can't carry much with them; but being of high metabolism, they need a lot of food for their weight. Their objective is not quantity per se, but staying alive, which depends on finding enough of the right food at the time it is needed. Call this the bird metaphor for the human objectives dealing with Compression.

Just from the outside, it appears that Toyota expanded too fast to sustain its original system for self similar scaling—the mentoring system. The objective is not to get bigger. It is to propagate an ever-improving process learning system. To do that, process scaling goes both ways, smaller as well as larger. Any communications network can become too big to communicate with itself, or too small to autocatalyze learning at the speed needed. Financial thinking seldom captures this because more money always seems better than less.

In human organizations, learners need learning time. Work is not all *doing*. Much of it is preparation for doing, or learning from an episode of work just finished, or absorbing the context for future work. Because learning takes energy, too, we have to make it more energy efficient.

Scientific Learning

Unfortunately, managers are apt to have a superficial understanding of scientific learning, regarding it as another gold mine, or a smart way to get more of everything, including a longer life—not as a way to learn faster. Rigorous scientific reasoning is counterintuitive to our natural approach to learning. We have to train our engrams to be more flexible, but disciplined.

The scientific method is traceable to Sir Francis Bacon's *Novum Organum* (1620).

Bacon posited low-level axioms, then tested them by documented observations. After confirming these, he posited higher-level axioms, testing them, and so on—rational determinism.[43] Gödel's theorem may have punctured rational determinism, but the scientific method remains the best bet to validate learning by checking whether reality as seen by multiple observers accords with a model of it.

Bacon soon found that human minds have many ways to duck this mental discipline. In archaic terminology he described "idols of human deception" that are still with us:

- Idols of the tribe (Bacon called all humanity a "tribe"): We see similarity where none exists and order in pure randomness. We rationalize wishful preferences and jump to conclusions rather than seek evidence, analyze it, and codify findings.
- Idols of the cave: We conform to our own culture, authority figures, and cool fads, rather than base our thinking on observed facts.
- Idols of the marketplace: Bacon's marketplace was of ideas, not goods or money (capitalism was in infancy). He meant that people are slow to accept new ideas, especially when advocates for the old ones remain very persuasive.
- Idols of the theater: We learn most vividly from dramatic anecdotes, received wisdom, and common sense, not from dull facts and causal logic. Theatrics also tempt popularity seekers to exaggerate their claims. Bacon spoofed William Gilbert, who discovered the lodestone (about twenty centuries after the Chinese).[44] Ever after he touted magnetism as the explanation for everything, much like exuberant consultants extolling the magic of proprietary techniques.[45]

When rigorously used, the scientific method *does not* claim absolute knowledge, but merely that any explanation accepted as a basis for further knowledge must agree with observations replicable by anyone else observing the same way. However, like everyone else, scientists are subject to compromising influences. The popular image of science emphasizes its outcomes, not its methods, reducing scientific evidence to just another marketing ploy, a commercial message laced with a few numbers and scientific jargon.

When money is at stake, scientific impartiality is easily bent. A discovery that favors a vested interest is a new marvel; if not, it's junk science. Scientists may not be pressured to recant facts, but they are certainly pressured to withhold them, or not to publicize them. Prominent examples are the decades-long public relations wars on the connection of smoking to health risks, and more recently the politicization of global warming.

Science is a predictive discipline based on empirically falsifiable facts, but evidence rarely stacks up as if bricking a wall. Later hypotheses may show earlier ones to be incomplete or, occasionally, flat wrong. Skeptics of any hypothesis must see evidence proportional to the degree of skepticism before accepting it.[46] Extraordinary claims require extraordinary proof. Explanations that agree with facts constitute science; without sound validation, impressive quantitative models remain conjectures. Insufficient validation remains very much an issue that sometimes raises a public hoohaw and sometimes not.[47]

The most rigorous descriptions of the modern scientific method are by skeptics devoted to debunking claims, scientific and otherwise.[48] Terms used in science differ from popular usage. For example, in science a theory is an explanation supported by considerable evidence, but in popular use, *theory* is often used derisively to mean *conjecture*. Even scientists abuse the term, as in "Theory of Venture Capital Suckage." A brief modern version of the scientific method in hard science is as follows:[49]

Observations lead to questions.
Questions lead to tentative answers, or hypotheses.
Answers are tested in a laboratory, in the classroom, or in the field.
Tests lead to modifications, and yet more tests.
Modifications ultimately lead to theories with supporting evidence.
Theories supported by overwhelming evidence lead to laws.

Just as in Francis Bacon's day, this is boring. Bias for action and making money enthralls far more people. Indeed, most undergraduates solve canned problems using accepted techniques, which poorly prepares them for lifelong discovery by observation and testing. They may even think that their first-learned techniques are absolute knowledge.

Focusing on results rather than the process that led to them obfuscates learning processes. Some decisions must be taken without taking time to collect and weigh evidence, so we get used to hip shooting. Even scientific research organizations may be inspired more by internal myths and folklore than fact, and working through human conflict issues entails as much psychology as establishment of fact, so becoming more scientific is no easy shift.

But without offsetting our instincts and clarifying cause and effect as best we can, we may not understand where we need to go, what has worked, what hasn't, and what unintended consequences may be lurking. In work vital to the world in the twenty-first century, that's becoming untenable. We need to learn and change much faster, rather than repeating the same old mistakes amid rancor and discord.

Scientific thinking is too important to confine to an educated elite. A much broader segment of the population has to learn to use a version of it every day in working organizations important to sustaining an industrial society's quality of life.

Self-Similar Scientific Process Learning at Work

Most basic research can only be done by scientists in a position to make observations impossible for most of us. We can't personally comprehend the complexity that took researchers years to learn or personally check all their evidence. We have to trust the integrity of their learning processes, no small issue in itself.

However, more prosaic human work is also done in complex, everchanging systems that no single person can understand in detail. For example, no one really knows how many individual parts are in a modern car (embedded in all subassemblies coming through several tiers of suppliers).

A number of stepwise methods based on scientific reasoning, like the Deming Circle, 8D, and DMAIC, have been applied in the business world. The labels on one of these frameworks are less important than rigor using any logic based on the scientific method.[50] Deming's Plan-Do-Check-Act (PDCA) framework shown in Figure 3.3 isn't easy to relate to, but that has an advantage. One is more likely to struggle to comprehend its logic rather than just plug through a cookbook procedure. Unfortunately, many companies water down this rigor to make training easier.

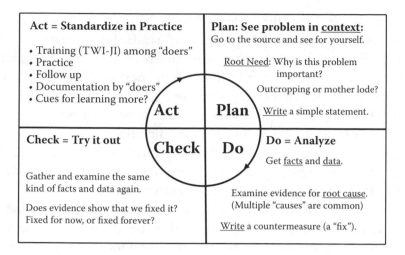

FIGURE 3.3
Brief but rigorous version of Deming Circle problem solving.

To use PDCA, people must be functionally literate and have a working grasp of a few basic quantitative methods. Although this is within the grasp of the average person in an industrial society, many intelligent, well-educated people have trouble rigorously carrying out a study by PDCA.[51] They are too impatient to carefully observe firsthand, too undisciplined to check how well a countermeasure works, or once proven, to hold it in practice. Eager to see results, they rush on to something else. As a consequence, they do not establish and preserve a base of learning as a platform for further learning. PDCA thinking is more like sticking to a diet and exercising regularly than binge eating followed by crash dieting. Like any other discipline, this requires overcoming impulses from human nature or from results-oriented business incentives.

To develop mental discipline, full-blown PDCA is excellent, and for more challenging problems it is necessary. Much of the time, however, less formal stepwise thinking will do: mentally ask five whys, try a fix, check if it worked, and standardize if it did, documenting as needed. In trivial cases, like where to put a trash can, just move the can and watch (marking a new standard spot if it's more convenient). But the key is to make this method for seeing and solving problems a daily habit. Then it applies in some form to every problem from positioning a pallet to developing an organizational strategy, a self-similar thinking process that scales up to macro issues and scales down to micro problems.

The first time around the circle (in Figure 3.3), start with the upper-right quadrant, PLAN, or with ACT if a previous standard exists, and work your way around. Once a countermeasure is validated (CHECK) and standardized in practice (ACT), start looking for the next problem with PLAN. Never stop the circle; it means that you've stopped learning.

PDCA is a general framework of thought. It does not specify a set of analysis tools. Some may be in wide use, and others may be highly specialized, depending on technology and the situation. It doesn't even favor induction or deduction. Its only preconception is to base conclusions on factual evidence using logic as sound as can be devised.

Sounds easy; it isn't. The self-discipline to actually complete going all the way around the circle is hard to acquire—but the potential is great if everyone in a work organization uses this thinking, if not always masterfully. In practice, work organizations have made great improvement in work processes using flawed versions of the Deming Circle and basic problem-solving tools. Many people rooting out many small problems either prevent bigger ones, or help simplify them.

A first impression is that applying this thinking to small-step problems might be great for repetitive work, like assembly lines, but not for creative work like market research. Not so. One of the big wastes is not establishing standards for routine work, and in lab research one can reduce experimental variance (and reruns) by establishing consistent lab techniques— standard work. But experimenters want to hurry on and get data to analyze. It is not just financially driven executives who become impatient for results.

To develop problem-solving discipline as teams, *everyone* in an organization should participate in exercises like full, disciplined PDCA until the thinking is habitual. Unless this thinking cycle is repeated again and again, and critiqued, habits do not form. Some of the most serious impediments are behavioral. People who collaborate seeing and overcoming problems have to learn how to get along together. One of the first behavioral habits to learn is to first investigate a work process before blaming a person. The way to form any habit is by practicing regularly, learning and improving processes by schedule, like Toyota (see Chapter 2).

Following the Deming Circle logic of Figure 3.3, process kaizen never has a stopping point. Devising a new method or making a technical discovery is just another step in learning, not an end point. Once it becomes a new standard or formally filed in organizational memory, it becomes a new platform for further learning. And a history of process changes is a knowledge bank to draw on when circumstances call for major changes. Process changes are then regarded more as routine than as a big disruption.

The ACT stage of the Deming Circle, standardization and documentation, is weakly executed in practice. Organizations rarely coach all employees how to properly train each other. Instead a training staff does it. But if processes are changed frequently, a central training staff is a bottleneck. Everyone has to learn how to train and how to learn, how to mentor and how to be mentored. And reward systems like individual bonuses should not motivate people to hold back what they know to look good by comparison.

Most of us also shortchange the PLAN stage. PLAN is habitually seeking and defining problems, constantly watching, sifting, and questioning work processes in the broadest context of which one is capable. PLAN implies deeply understanding a process and its possible consequences far removed in both time and distance, in addition to nibbling away on small improvements in what is done now, until even questioning whether a process can be eliminated or its purpose served in an entirely different way.[52] But the key to making the Deming Circle a continuous circle is making sure process learning is a regular activity to be scheduled just as musicians schedule practice time, both as individuals and as an organized group.

Broadening Your View of Waste

Whether we are engaged in small-step process improvement or big-step innovation, how do we know we are making headway on the challenges of Compression? One way is by using measurements like those in Figure 3.4, which is a chart similar to one in Chapter 2 and is useful for evaluating both big-step and small-step changes. Specifics vary, but the logic and measurement approaches for process Compression are self-similar from small scale to large scale.

Big-step, radical innovations may not occur as regularly as improvements of existing processes, but they should not be rare events inspired only by genius either, as noted by Thomas Edison, whose laboratory systematically ground out innovation after innovation. For his time, Edison's factory commanded a broad scope of technology (context), plus manpower to persist in trial and error. Finally hitting on a superior light bulb filament was really just a process improvement exercise with a bigger impact than

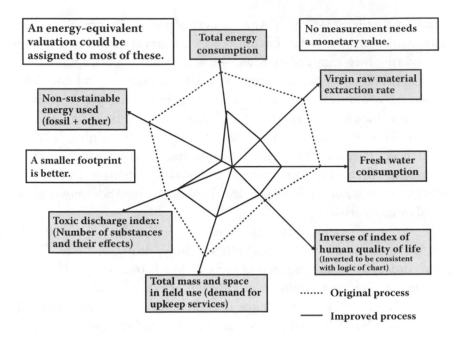

FIGURE 3.4

Multidimensional process performance comparison for both big-step and small-step process compression.

most. Besides, Edison's real innovation was a complete lighting system, a package of inventions. The light bulb only symbolized them.

Figure 3.4 is a generic template, not a real case, illustrating how the same thinking that guides process kaizen, as discussed in Chapter 2, can be expanded to estimate the reduction of environmental waste. If the measures are scaled so that a smaller value always represents improved performance, shrinkage of the area under the dotted lines becomes a visual indicator of the degree of process improvement. No measurements are in dollars or any other currency. All are physical process indicators: a unit count or total mass, volumes like "space consumed," or ratios of physical measurement indicators. All measures are interpreted by humans, of course, but attaching human valuations to them can warp the measures. If a change is a genuine improvement, waste (or total resources used) should decrease. A real improvement (Compression) should show a gain by at least one indicator, with no accompanying decrements by any other indicators (no trade-offs).

All the measures plotted in Figure 3.4 are physical measures of process performance, so they are the same anywhere in the world. Nothing is converted to currency. Mapping the process changes into dollars or any other currency helps us estimate how they will play in a business model, but if we're seriously addressing Compression, we must recognize that the financial system is a human system. It does not directly measure real, physical changes. Therefore, to evaluate physical process changes, performance measures should not be more degrees of abstraction removed from physical reality than necessary.

To those accustomed to finance and accounting, this seems strange. All other considerations seem subordinate to our financial survival, and surely anything important can be tied to financial measures. But the transactional world is an artificial human construct, and the assumptions in this thinking are loaded with self-deception.

One self-deception is the economic orthodoxy of trading off use of one resource for another. Certainly that has to be done, just as preindustrial farmers always had to set aside ground to raise next year's seed and to feed animals. This logic is the heart of capitalism: that making an investment demands making a return. But it makes it tough to grasp the heart of Figure 3.4. True process improvement decreases the use of all resources to accomplish the same thing—or better.

This is outside of conventional business thinking. If you put in nothing except human ingenuity, but get a lot back, the rate of return is infinite. Of course, from a financial view, this makes no sense, but dealing with Compression goes outside the assumptions that financial models were designed for—expansion and substitution.

Figure 3.4 shows process measurements similar to those used for lean conversion of operational processes. Lean process measurements guide us toward conserving resources, but they are not enough to deal with Compression.[53] The lean view of the system is too small. Lean conversions generally begin with a segment of a total life-cycle process, perhaps just part of a value stream inside one factory, and eventually it may extend to all the processes used by suppliers, and those used by customers, but rarely does it cover the full life cycle of a product or service. In a global market economy, tracking this is almost impossible, and as recent problems with Chinese imports into the United States reveal, a huge problem. In a preindustrial economy, on an old farm, the producer and consumer were frequently the same person.

A local producer/consumer could directly see, to some extent, the effects of her stewardship on the natural resources nearby. For example, she could easily see that land was eroding. In a global industrial world, it is much harder to measure the effects of our processes on global, physical reality. But before going on, once again, the goals of dealing with Compression are as follows: Assure survival of life and promote quality of life using processes that work to perfection with self-correcting, self-learning systems. No use of excess resources. No wasted energy. No toxic releases. Quality over quantity, always.

Arbitrary quantification sharpens the impact of this wordy statement: Worldwide, create at least the same quality of life as in industrial societies today, while using less than half the energy and virgin raw materials, and cutting toxic releases to nearly zero. So how can we know whether we are moving in this direction?

Mapping Your Opportunities

Companies that are familiar with lean manufacturing and value stream maps may use them to identify opportunities for resource conservation or mitigation of toxic releases. A rough example is shown in Figure 3.5.

The abbreviated value stream map in Figure 3.5 doesn't even show inventory points. It's kept simple to illustrate the notes on Compression opportunities associated with each operation. This helps lean practitioners start with a tool they know how to use, but it is not enough. They have to see the huge scope of the challenges.

A technique to help expand the scope of practical thinking is shown in Figure 3.6. It's a mass–energy balance, a diagram of everything that goes

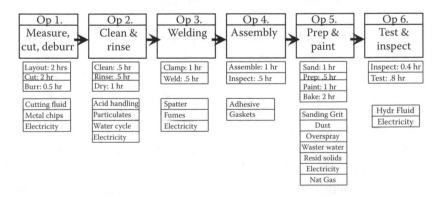

FIGURE 3.5
Value stream map used to identify larger-scope waste opportunities.

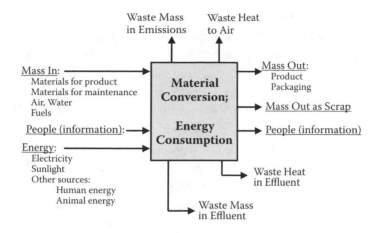

FIGURE 3.6

Diagram of a mass–energy balance. Based on diagrams used by Dave Gustashaw at Interface Inc. With permission.

into a system, and everything that comes out. Variations of this have been used for everything from design of oil refineries to monitoring the nutritional input–output of medical patients.

An input–output diagram as in Figure 3.6 can only be done for a system having defined boundaries, but they do not have to be walls. For example, such analyses have estimated input–output across the boundaries of ecological zones. Figure 3.6 presumes a factory, or part of one, converting material into product; but input–output analysis can also apply to sales offices, call centers, IT centers, clinics, police stations, airports—the physical nature of any operation.

For any system, diagramming the mass in and out, and the energy in and out, expands the picture measured by financial transactions in dollars. Those see only inputs that are paid for and outputs that can be sold. However, inputs and outputs in Figure 3.6 would be in pounds of mass and BTUs of energy (in the English system). With the exception of chemical plants, few work systems perform this kind of input–output analysis today. In chemical systems, the properties of outputs may be very different from those of inputs, so a mass–energy balance needs to account for chemical reactions and material phase changes. That makes it get technical, but input–output analysis should be simpler for most process systems. It's a way for every work organization to more clearly understand what it takes from nature and what it discharges to it.

A mass–energy balance is more holistic than the value stream maps usually used for lean operations. Those often map only the flow of materials presumed to represent the value added to customers. They seldom include more than the movement of people, information, and money. By contrast a mass–energy balance represents *everything* going in and everything coming out of a process, no matter how humans value it. Every molecule going in and out will never be detected, but without giving it thought, much that could be seen is neglected. That is the point. Measure *all* resources used in a process and *all* outcomes from it, not just the ones that humans value.

Input–output analysis of a restaurant, a farm, a dry cleaner, or a swimming pool can be eye-opening. For example, part of input–output analysis for a restaurant would include checking what goes into the garbage, which can uncover many ideas to both reduce waste and improve customer satisfaction. (Restaurant chains have done garbage can audits for decades.) In practice too much input–output analysis is incomplete. All the outputs don't add up to all the inputs, so that it really does balance.

By expansionist economics, limiting vision makes sense; why look at something that has no value? Holistic input–output analysis differs from economic orthodoxy because it forces us to understand what we do, not just what we will get. But to more fully comprehend what our human-controlled processes take from nature and give back to it, we need to expand the system boundary for mass–energy balance both geographically and in the time span covered.

Life-Cycle Analysis

Life-cycle analysis usually refers to the entire life cycle of all the materials in a physical product. Most nondegradable products have a birth-to-death cycle, from ore in the ground to junk in a landfill. If the product consumes energy and materials in use, that adds to any life-cycle mass–energy balance. In a few cases, companies have a cradle-to-cradle life cycle. Materials are fully reclaimed and reused. The objective in Compression is to compress the mass–energy footprint of a complete usage cycle while preventing toxic releases at the same time.

Considering its importance, our competence doing life-cycle analyses needs considerable improvement. Most of us can only trust the calculations of experts, presuming that they have data and have accounted for a full mass–energy balance estimate. The holes in these analyses make the news now and then. For example, a plug-in hybrid vehicle stretches fuel economy by the engine, but energy at the plug must come from somewhere, and then one must factor in the extra energy and materials for a full life cycle of the battery. We need to collectively learn a lot more about this in a hurry. One of the places to make headway is improving the relevant measurements.

Practicalities of Measurement

To do a mass–energy balance on any system, *all* inputs and outputs may be difficult to measure. Some are unknown, or if known, means to measure them are not at hand. Given that, it's better to measure things that are important approximately than unimportant things precisely. Life-cycle analysis and mass–energy balances for environmental sustainability try to measure the effects of human processes on the big, open system called nature that lies *outside* them.

Approximate measurements, or even best guesses, are better than convenient measurements, like the old cliché of a drunk man seeking his lost car keys under a lamppost because that is where the light is. For example, thermodynamic measurements of energy from natural sources, or which is escaping as waste heat, may be rough approximations. Engineers frequently estimate energy changes from specific heats, the energy required to elevate a unit of mass of a given material by a unit of temperature. Such measurements help estimate whether the energy expended per drink is greater using reusable glass bottles or recycled aluminum cans.

A related, but neglected measurement is Howard Odum's "emergy."[54] Sloppily defined, *emergy* is the cumulative energy that has transformed a material or system into its current state. A system may be a small bit of material or a big ecological area. Ecologists use emergy analyses to estimate whether, over some time period, an ecological area has gained or lost energy. A big energy imbalance—big gain or big loss—indicates an ecology in transition, which is not sustainable because a big energy imbalance must correct sooner or later. On earth, natural processes distribute energy around many complex circuits feeding back on one another, maintaining balance. A regional ecology exists in an ever-shifting thermodynamic equilibrium of weather and complex biochemistry with the whole earth. Overall, the whole earth is not efficient; it just "is," but the thermodynamic balance of regions indicates how closely they are in balance with the whole.[55] Within them, individual organisms become extraordinarily efficient at capturing wisps of energy to perpetuate the species. Ecologists are almost the only people that use emergy in analyses, but the concept is potentially valuable for many of the rest of us.

Hydrocarbons accumulate energy underground over geological lengths of time, attaining exceedingly high emergy by a natural process. No living organism has the high energy density of petroleum. Plants capture solar BTUs beamed over many seasons. As these slowly compact under pressure they form hydrocarbons—high-density natural energy storage. Burning them dumps this lengthy accumulation of energy and CO_2 into the ecology. Burning recent growth taps a brief accumulation of energy—within the cycles of energy replacement from the sun. A sustainable system keeps a better ecological balance by replenishing energy used with energy freshly generated from the sun—and a few other sources—without unbalancing the system of replenishment.

Emergy is a bank of nature's dollars, good anywhere in the universe, and independent of human values. Nature does not value human currency. It only responds to what humans do to it physically—trades only in its own thermodynamic currency, which translates to human financial accounting only if energy exchange involves a money transaction. Odum sometimes transposed joules of energy into "em-dollars" to help financial addicts relate to this.

Mass–energy input–output analyses over a wide-scope system are needed to determine, for instance, life-cycle energy savings from LED lighting or fully electric cars. The total energy actually consumed in a product life cycle must include all the energy to find and refine all the raw materials, transport them, convert them—all the way to the energy to dispose of them, or recycle or reuse them. Even environmental appraisals of alternative fuels often fail to consider all this, but assume that they are acceptable if a human market will pay for them.

Input–output analysis and life-cycle analysis has to include "green chemistry"—be qualitative as well as quantitative. Qualitative analysis is crucial to evaluate whether the molecules either required for a process or emitted from it have bad consequences. When released into nature in even small quantities, determining their potential effects long hence can uncover unpleasant surprises, as illustrated by the story of Freon and other halogen gases and the ozone hole. Data may not be easy to come by, but this is not a matter to shrug off.

Much is yet to be discovered or confirmed about biochemical contamination. Effects can be subtle, unexpected, and long-delayed in appearance.

For example, evidence suggests that not only is male fertility dropping in all industrial societies, but some evidence shows a feminization of males in many animals. The consequences are serious enough to merit thorough research.[56] Few of today's companies are set up to dig deeply into this new layer of complexity. That is why zero release of any chemical not sure to be benign is a good goal.

This kind of thinking has to extend to design for the environment (DfE). Computer-aided design with design rule prompts, simulation, and reference database checks are possible, subject to the usual caveats about models. Space does not permit an extensive review of DfE, but it is a highly integrative process, sometimes called cradle-to-cradle design.[57]

The measurement of environmentally sustainable processes is far from mature. Three well-known examples of a proliferation of performance criteria are the EPA's environmental management system performance track,[58] McDonough-Braungart Design Chemistry's (MBDC) cradle-to-cradle certification rating,[59] and Walmart's packaging scorecard for suppliers, all 61,000 of them.[60] None is truly comprehensive.

Another issue is finding enough data to perform DfE knowledgeably or for auditing any complex DfE case. Process owners are reluctant to disclose process data in open sources for "national security or competitive reasons." This secretiveness is a big problem. Paul Chalmer of the National Center for the Manufacturing Sciences has a project to construct a base case of the energy and other factors involved in producing some common materials that may be used in product design. It's a start. There's a long way to go.

Action Steps to Start Learning

A short, simplistic list of steps to develop environmentally sustainable processes is as follows:

- Reduce the mass of things produced, moved, and used; and reduce the space they consume—quality over quantity always.
- Reduce the total energy consumed in processes by which they are made, moved, stored, and used; and improve the yield from the energy actually used.
- Create processes that reuse, remanufacture, recycle, refurbish, and upgrade, driving the need to use virgin material from the earth to a minimum.

- Eliminate the need for processes that release chemicals known to be toxic to any form of life. If they cannot be eliminated, sequester them from the environment.
- Promote the use of materials and processes that enhance the productivity of natural solar energy cycles and biological self-regulation.

PART 3: ORGANIZATIONAL LEARNING

The mass of detail to be addressed makes it impossible to cope with Compression by forcing work organizations to do it through regulation. That's slow feedback learning, accompanied by foot dragging and procedural waste. Work organizations have to understand their mission and objectives dealing with Compression, and want to fulfill them, with everyone autonomously working to that end. These vigorous learning organizations need disciplined learning systems, consisting of common tool sets and collective "lab notebooks" embodying a *de facto* learning language. But most of all, they must institute the behavior for disciplined learning; otherwise we can never overcome the human weaknesses outlined earlier in this chapter.

The frontier for process learning is in the entrepreneurial end of it, low self-similarity, learning where innovation is the most deviant but fear of unintended consequences is high. Every loner with a hot idea cannot do anything they want when unanticipated consequences can be drastic. We need to harness entrepreneurial spirit differently, creating twenty-first-century innovation factories that cope with a lot more complexity than a hodgepodge of incubators, venture capital networks, review panels, university technology transfer programs, and big company "intraprenuership" methods. We also need to get beyond the market test as the primary learning arbiter.

To make this shift, the primary goals of all work organizations essential for human welfare need to shift: from making money to ultimate performance, from concentrating on what the organization *gets* for us to what it must *do* for us. This is a radical change. At our best we cannot do it quickly and easily. Here are some thoughts for organizational leaders of this change on how to start:

Establish a social mission for the organization, including some shorter-term goals for better performing it. Involve people working in the organization in this so that they actually think through the *why*, not just comply.

Expect all people to become professional problem solvers, both in attitude and in skill. Lead by asking questions. Expect everyone to think. Introduce a scientific framework of thought like PDCA, and some common tools to apply using the framework, including mass-energy analysis. Stimulate people to actually practice problem solving regularly—even by a schedule.

Create learning cycles and make them shorter and shorter by continuously improving the learning processes themselves. Regard learning as part of any other work process, and strive to improve it in a similar way.

Develop the visibility of work and work processes to prompt as much learning attention as possible. Develop people to train and mentor each other so that improvements or findings are held in common as the actual base for the next advancement. Devolve responsibility onto teams to manage themselves and their processes to the maximum extent that they are capable.

Glue this together with a common problem-solving language and data-bases, based on concepts like PDCA, A3 papers (see Figure 3.7), and search systems as easy to use as Wikipedia. Integrate knowledge manufacturing into all work. Make it high priority.

Expand the context for learning, beginning with increasing visibility of work at the scene, and expanding this to networks of learning organizations. In the current context of supply chain management, this is sometimes called an extended enterprise, but we have to get past dependence on business language. Open up systems, so there are no secrets for the core workforce, including financial information, and there is little or no difference between organizations because of the source of financing. Secrecy has to abate by diminishing the learning speed bumps of intellectual capital monopolies.

Coach people in efficient communication, in meetings, by e-mail, and by using other media. Coach them in listening—in real dialogue examining facts and testing logic—not in contending for a win. That is no less than an initiative to civilize normal, tribal human behavior, which may be our biggest human challenge.

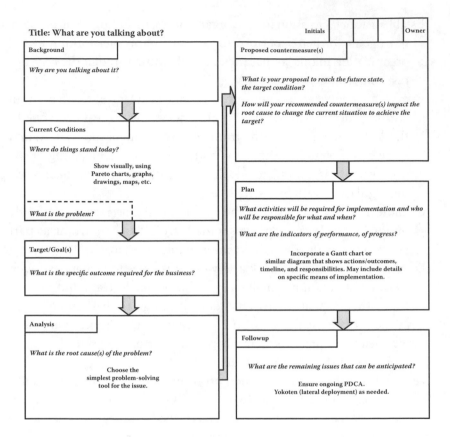

FIGURE 3.7
Typical format for a one-page A3. Used with permission of John Shook, *Managing to Learn*, pp. 8–9.

Creating a Learning Language

In recent years knowledge management systems have become popular to help people find people or reports inside the company that can answer a question. Knowledge management is usually limited to creating better searchable databases, better benchmarking exchanges, and more networking within "communities of practice."[61]

As a graduate student 40 years ago, I worked in a technical information service, a crude punched-card forerunner of Google. Then, as now, the heart of the service was helping clients clarify their problems, and then matching them with knowledgeable experts that could personally assist them. Translating a problem into a common phraseology was frequently the key. That is, regardless of technology, problem solving is a human activity.

Today, the Internet has many all-purpose search engines, but ineffective if we don't know either sources to search or the language of the inquiry. For example, if we're looking for medical information, but don't know the terms, finding a valid source can be quite a learning process, and charlatans may prey on our ignorance.

In working organizations, learning-related confusion tends to be vaguely classified as communication problems, almost always topping the list of employee gripes and stirring a range emotions they find difficult to pin to any specific cause. We don't like being in the dark. We fear looking stupid asking questions. A learning system has to clarify this mush.

Thinking of this as an information system problem, the key to faster learning is better codification of what we know. Figure 3.7 shows one of the better-known ways that work organizations are doing that today. It's an A3 paper modeled after those used in Toyota, and many variations of this format exist, both in Toyota and in other organizations. They are effective as long as users share a common basis for defining a problem, usually the case if the mission is as clear as "design, test, produce, sell, and service vehicles."

The purpose of an A3-like paper is to clarify thinking. The more that problem logic and countermeasures can be presented in a self-similar format, the easier it is for others to decipher what happened, what was done, and why. Note that the A3 format in Figure 3.7 presses authors to clarify all key aspects of their thinking to others in a small space. Consequently, A3 papers are a medium to improve learning and dispel some of the fuzziness of poor communication.

Like PDCA, an A3-type format is self-similar to scientific problem solving, so it promotes this language at work. If they are well created, the human process of using them should invite readers of A3 papers to personally contact authors—go to the source. Developing other communications media with the syntax of scientific thinking will further promote this language until it finally begins to constantly reinforce itself—becomes embedded in the working culture, which is the real intent, not the adoption of some magic technique.[62]

Disclosing knowledge of failures is just as important as disclosing successes, so old failures are not repeated. Fresh thinking or new findings may turn them into successes.

Such openness extends the principle of visibility to every part of an organization, including its network of partners and contributors. Visibility is unwritten, unspoken communication that eliminates the waste of communicating the obvious, or the emotion of asking dumb questions. Routine operations run themselves, so that attention can go toward second- and third-loop learning, and to problems that people had no time to pay attention to before.

Rapid learning comes out of free energy density, concentrating knowledge where people can see a lot and find out a lot very quickly. Extend this principle to organizational design in research and test labs, studios, marketing offices, call centers, and IT programming cubicles. Set up war rooms, closed-circuit video cams, or any other medium that will help, while not overloading limited-capacity humans at the node points in such systems, which appears to be the major limiting factor.

Many tools to assist this exist now, ranging from Kepner-Tregoe to TRIZ (Russian acronym for *theory of inventive problem solving*).[63] Some may be unique to an enterprise's technology. Others are helpful in a broader context when people don't agree on objectives or whether a problem is real; that is, they have a wicked problem. Examples are after-action reviews and the SOL coaching model.[64] The number of such techniques is unknown, and probably unknowable. As long as they are self-similar to scientific thinking, the exact tools are less important than weaving them into a rigorously used organizational language for problem seeing and problem solving.

Innovative Learning

Asked to name an innovative organization in 2007, one is apt to think of companies rolling out hit products, like Apple Computer. A Toyota is likely to be far down the list. Basic automotive technology is seen as mature and therefore migrating slowly. By Charles Fine's Clockspeed rating system, aircraft builders have an even slower new-product cycle than automotive.[65] In both cases, very advanced technology goes into new designs, but the basic product concept is old, systems are complex, and reliability is paramount. System integration and testing take time. Process innovations are less publically visible. For example, Walmart's cross-dock logistics and speedy checkout innovations helped give it a cost edge, which is what people see.

True innovation changes what we do. Ideas that never materialize are a dime a dozen, and mere inventions may be tucked on a shelf, never actually used. A discovery is of something that exists, but was unknown, like genes. Machines that allowed genes to be discovered were invented. When they are used for routine genetic identification typing, they enable innovation, people doing something very different. Much childhood learning is rediscovery of things that adults already know—but once in a while, some things that adults don't know.

When products and processes change, radical innovation breaks up a rigid organization. Systems, status, and the business model ossify in support of old technology; when a big move is called for, old bones may not loosen up fast enough to keep them from cracking. The old story of Xerox giving iMac technology to Apple is a good example. Xerox could not risk marketing a truly new concept, but Apple, with minimal historical baggage, could grow a new organization from a simple base. Economist Schumpeter famously dubbed commercial innovation as "creative destruction."

Organizational innovation can be compared with punctuated equilibrium in evolutionary theory. As an ecology evolves, change becomes less rapid until it stabilizes in some near-steady state. If a major event collapses this existing order, it has to start over, eventually reestablishing a new ecological balance.[66]

When we think of radical commercial innovation today, it's seldom of Toyota, Boeing Commercial, or the American Electric Power Institute (research arm of the power industry). We think of the latest coolware from the MIT Media Lab or Silicon Valley, where popular impression of the business model is venture-capital roulette: Bet on ten to get a single hit, winners paying for losers. Intuitive gambles on fickle markets—movies, music, fashion—are like that. So are gambles on acceptance of radically new technology. (Reality is that financiers who are "betting on the come" back some flyers for years.) By contrast, innovation by work organizations that must get it right entails much more testing and learning. Reality also is that any innovative organization contending with the challenges of Compression has to preclude much more risk. Consequences can be much more serious than whether a movie flops. An organization that must innovate while dealing with Compression needs a learning model conceptually like Figure 3.8. If innovation requires a change in all aspects of operations because it is a business model change, it has to be able to step up to that, too. All kinds of ideas can emerge from science, technology, and perhaps even needs for serving customers. Either discarding them as faulty or reducing them to faultless practice is really a learning process that rapidly perfects the processes for doing this.

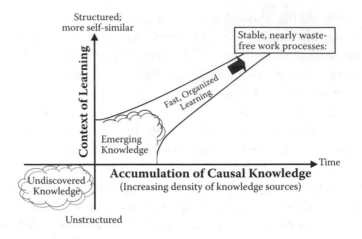

FIGURE 3.8
Compression of organizational learning processes.

There's a lot to learn very fast. Instead of an established organization, consider an entrepreneur trying to turn a dream into reality. If inexperienced in business today, she has to learn about persuasion, law, permits, financing, business plans, payrolls, false rumors, marketing, customers, leadership, and many other things. Flaws in the original dream need to be worked out, and even the original concept of customers for an idea may undergo revision. If this education process drags on for years, entrepreneurs not manically obsessed give up.

Such entrepreneurship is frequently a community learning project today if an entrepreneur is willing to be coached. To attract capital, the entrepreneur must usually split the take with others that bet on her, so one early lesson typically is that a percentage of any long-term return beats 100 percent of zero. But this move also transforms her baby into a community child with many godparents that want it to succeed. If she has wise mentors and heeds them, her venture has a better chance of survival.

This kind of learning process can always use improvement, and it helps if members recruited to a learning team focus on what they must do more than on any fortune they might get. A learning team reduces the entrepreneur's imaginary waste (from ignorance).

Silicon Valley has many support mechanisms for startups besides venture capitalists—contract technicians, prototype shops, legal services, and the like—so ventures are apt to become enterprisewide learning projects. If *everything* necessary to hit a tight market window with a hot idea has to be learned from scratch by novices, the project is more likely to be botched. Because this kind of learning support structure encourages transferring novel ideas from universities or labs into commercialization, communities outside Silicon Valley have tried to replicate it for high-tech expansion.

This kind of system sets up tension between open-system learning (more an academic model) and competitive learning (commercial model). Competition racing to market riches is instinctively motivating, and the core of expansionist thinking. But in Compression even startups will have to consider their impact on resource use and the global environment and check for unintended consequences. Making the public guinea pigs for poorly thought-out ventures is an approach that has to go. And enthusiastic entrepreneurs think that they are repressed by regulation now.

In this new world, reality is that no venture is a limited liability organization, no matter what its legal charter says. Everybody has to become more socially responsible. In this light, organizational learning also has to become more collaborative than at present (see Figure 3.9). The motivation for decreasing the time of learning is less to hit a jackpot than to get everything right—quality over quantity.

The contrasting Johari windows in Figure 3.9 show the direction that more innovative learning has to go—from a competitive model more toward a collaborative one.[67] Competition naturally stirs more adrenalin than collaboration, and even academic researchers can be intensely competitive and have notorious tiffs; but in the end, getting it right is what has to count the most.

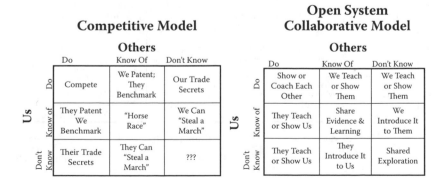

FIGURE 3.9
Competitive versus collaborative organizational learning.

Forming Bonds of Trust

Forming an innovative team or a startup team frequently throws a lot of relative strangers together. Having a common process learning language is not enough if they don't trust each other deep down. Diverse people do work together when they do not even like each other very much, but they have to keep personal conflict from getting in the way. They have to trust each other for work purposes.

Douglas Smith's bond thickness, summarized in Figure 3.10, is one way to describe trust relationships in human organizations.[68] Trust is critical—trust that others will at least try to do what they say. A group with thick bonds can self-organize for different purposes, the ultimate in collaborating, but even strangers meeting in the thin bond of a market must trust that other parties will pay up. Without that minimal trust, there is no market.

People who know each other in multiple contexts and through multiple activities have high bond thickness. If they interact only in one context, they have low bond thickness. However, interaction does not automatically confer trust. Enemies may know each other well. People living in communities are more likely to form "thick we" relationships because they have more interaction opportunities; however, a "thick we" comes from success in doing things together—mutual dependence. A military combat

		\multicolumn{5}{c}{**Basis for Human Organization**}
		Market	Network	Working Organization	Family	Tiger Team
Basis of Commitment	Motivation	Mutual $ Gain	Mutual Learning	Rewards, Mutually Dependent Activities	Mutual Traits, Filial Bonds	Mutual Victory, Mutual Survival
	Communication	Prices & Specifications	Information, Contacts	Why-to, What-to, & How-to-do	Personal Knowledge	One-for-All; All-for-One
	Durability	Minutes to Months	Months to Years	Months to Years	Lifetime; Years	From Days to a Lifetime

←———— **Bond Density** ————→

Weak ("Thin We")　　　　　　　　　　　　Strong ("Thick We")

FIGURE 3.10
Organization types and human bond density.

unit or any other group in which lives depend on each other has to form thick bonds very quickly.

Bonding thickness is important for sizing work organizations. No human can form a "thick we" with six billion other people on earth. Small tribal communities bond naturally, but are usually isolated from a wider world. By contrast, a global working organization is a community of purpose, but many people never meet face-to-face. If a work group's turnover is high, bonding is thin and temporary: here today; gone tomorrow. Indeed, bonding may be thicker between alumni of work organizations than between people now working in them.

All this bears on the right size for a work organization capable of addressing problems that may be global in scope. To build the trust necessary for open communication and shared learning, key players in a geographically scattered work group need a bond thicker than people trading cars. As many managers have learned, they need to meet face-to-face, see body language, and sense each others' character and capabilities before they can truly open up by phone and computer.

Higher Context Learning

What is *context*? When working by computer, can you see errors either in text or in graphics more easily when printed on paper than when on the screen? Most people say yes. Screens narrow the scope of perception, which is a loss of context. In a broader sense, high-context learning means learning about the whys of the work, what seems to really be needed, and what other people do.

Context learning has also been used to refer to educating people from high-context cultures in low-context settings.[69] In a high-context class, people may sit in a circle sharing conversation, perhaps with common artifacts on display; but in a low-context class, students sit auditorium style, and most exchanges are with the teacher. Higher-context people may perform poorly in this setting.

Low-context work training occurs before workers go into the workplace. High-context training is more apt to be embedded in the work, with more stress on why work is done and how it relates to what other people do. Low-context organizations focus on getting something accomplished. Higher-context ones (like Toyota) focus more on how and why work is done—on work processes and developing people. If work must consider environmental issues far removed from the immediate scene, organizational learning has to become higher-context. For example, engineers need deep knowledge of some specialty and equally broad knowledge of other considerations that put it into context. Indeed, designing for both innovation and the environment is as much about integrating a huge number of diverse considerations as about advancing technology. Encompassing so much detail exceeds the capacity of a single human mind. That's why higher-context organizational learning systems are necessary.

Indeed the challenges of Compression seem to exceed the capabilities inherent in the ways we organized work during the twentieth century. We need something even better than the lessons absorbed from that twentieth-century icon Toyota.

ENDNOTES

Extended version of endnotes available at http://www.productivitypress.com/compression/footnotes.pdf.

1. Inability to promulgate systemic thinking frustrated its pioneers: Stafford Beer, Henry Minzberg, Jay Forrester, and others.
2. Chris Aryris and Donald Schon, *Theory in Practice*, Jossey-Bass, San Francisco, 1974.
3. Bateson, best known as the husband of Margaret Meade, was a distinguished anthropologist on his own.
4. Donald O. Hebb, *The Organization of Behavior*, John Wiley & Sons, New York, 1949.
5. From Gerald M. Edelman, *Second Nature: Brain Science and Human Knowledge*, Yale University Press, New Haven, 2006, p. 23. Edelman attributes them to J.B.S. Haldane.
6. Ivan Amato, *Super Vision*, Harry N. Abrams, Inc., New York, 2003 (opening chapter).
7. Eye physiology sources include Amato, op. cit.; and http://webvision.med.utah.edu/facts.html.
8. William Sheehan and Thomas A. Dobbins, "Lowell and the Spokes of Venus," *Sky & Telescope*, July 2002, pp. 99–103.
9. The Harvard site for this video is gone, but a similar video has been produced by Viscog Productions.
10. Gary Giddons, "Put Your Voice Where Your Mouth Is," *New York Times*, Dec. 29, 2004.
11. Susan Blackmore, "Abduction by Aliens or Sleep Paralysis?" in *Bizarre Cases*, (CSICOP), Amherst, NY, 2000, pp. 30–39.
12. Mark Ritchie, *Spirit of the Rainforest*, Island Lake Press, Chicago, 1996.
13. Bertrand Russell and Alfred North Whitehead completed *Principia Mathematica* 1913, attempting to construct a "mathematical theory of everything," a major ambition of early twentieth-century logicians.
14. Alan Turing (in 1937) and Greg Chaitin (much later) proved that to solve some propositions, a computer would never stop running without deciding anything.

15. Douglas M. Hoffstadter, *Gödel, Escher, Bach, an Eternal Golden Braid*, Random House, New York, 1979.

16. An English version of Gödel's theorem is at http://home.ddc.net/ygg/etext/godel/godel3.htm.

17. David Berlinski, *Infinite Ascent: A Short History of Mathematics*, Modern Library (Random House), New York, 2005. In non-mathematical language, Chapter 9 summarizes the history and significance of Gödel's theorem.

18. Children unable to grasp abstraction learn the fable version: A lazy hare wakes up too late to overtake a tortoise.

19. In the 19th Koan of Mumonkan at http://www.angelfire.com/electronic/awakening101/mumonkan.html.

20. Rebecca Goldstein, *Incompleteness*, Atlas Books (Norton), New York, 2005.

21. Douglas Hofstadter, *I Am a Strange Loop*, Perseus Books, New York, 2007 (on symbolic self-referencing).

22. Probably the best-known antirationalist philosopher was Jacques Derrida, famed for "deconstruction."

23. Howard Gardner, *Frames of Mind: The Theory of Multiple Intelligences*, Basic Books, New York, 1983.

24. All the early cyberneticists came to see that a corporation must be more organic in structure than a strict hierarchy, for example Russell Ackoff, *The Democratic Corporation*, Oxford University Press, 1994.

25. Philip E. Ross, "The Expert Mind," *Scientific American*, August 2006.

26. Elkhonon Goldberg, *The Wisdom Paradox*, Gotham Books, New York, 2005.

27. "Those Who Were Once Blind Can Learn to See, Study Shows," *Science Daily*, Feb. 20, 2007.

28. Jeff Hawkins (with Sandra Blakeslee), *On Intelligence*, Times Books, Henry Holt, New York, 2004.

29. Maia Szalavitz, "So Long Crystal Meth, Hello Cocaine," *Statistical Assessment Service* (www.stats.org) June 12, 2007.

30. Charles Taylor, "Blind Spot: Hitler's Secretary," *Salon.com*, Jan. 31, 2003.

31. Susan Choi, "The Bubble Girl," *New York Times*, July 12, 2004.

32. William Perdue, *Terrorism and the State*, Praeger Publishers, Westport, CT, 1989 (written long before 9/11).

33. Search "water chemistry" or "water physics" on the *Science* site: http://www.sciencemag.org/.

34. Clayton Christiansen, *The Innovators Dilemma*, Harvard Business School Press, Boston, 1997.

35. T.P. Wright, as reported in "Factors Affecting the Cost of Airplanes," *Journal of the Aeronautical Sciences*, February, 1936.

36. Zipf's law was developed by experimentation and observation by George Zipf, a Harvard linguist.

37. A syntax pattern is sometimes illustrated by asking people to count the number of letters in messages (with lots of errors made). Once our engrams detect what they expect to see, we stop processing symbols in detail.

38. Noam Chomsky, *Aspects of the Theory of Syntax*, MIT Press, Cambridge, MA, 1965.

39. Eric Chaisson, *Cosmic Evolution: The Rise of Complexity in Nature*, Harvard University Press, Cambridge, MA, 2001.

40. Stuart A. Kauffman, *Reinventing the Sacred*, Basic Books, New York, 2008.
41. For example, R-Theta of Mississauga, Ontario, does this on heat sinks for electronic equipment.
42. Hans Baumann, *The Ideal Enterprise*, Vantage Press, New York, 2002.
43. Bacon codified the thinking of other scientists of his time. Whether he personally "did much science" is doubted.
44. Robert Temple, *The Genius of China*, 1986 (Carlton, London, 1998).
45. To give William Gilbert his due, despite his exaggerated claims, he advocated evidentiary reasoning *before* Bacon wrote on the subject. Even Isaac Newton dabbled in alchemy in addition to calculus and physics.
46. The philosophy of science covers the logic and rules of scientific evidence. A readable modern explanation is Carl Sagan, *The Demon-Haunted World: Science as a Candle in the Dark*, Ballantine Books, New York, 1996.
47. Orrin H. Pilkey and Linda Pilkey-Jarvis, *Useless Arithmetic*, Columbia University Press, New York, 2007.
48. K.D. Bomben, "Scientific Truth," used as a reference by Skeptic Friends Network.
49. Also from K.D. Bomben.
50. W. Edwards Deming extended a similar schema of his mentor, Walter Shewhart. See John Butman, *Juran: A Lifetime of Influence,* John Wiley & Sons, New York, 1997.
51. The most difficult challenge of the author's teaching career was coaching college juniors to actually use PDCA.
52. This isn't new: E. Bright Wilson's description of the scientific method in Chapter 3 of *An Introduction to Scientific Research*, McGraw-Hill, New York, 1952, stressed careful observation as the key to original discovery.
53. Gary G. Bergmiller, *Lean Manufacturers Transcendence to Green Manufacturing: Correlating the Diffusion of Lean and Green Manufacturing Systems*, Ph.D. dissertation, University of South Florida, 2006.
54. Howard T. Odum, *Environmental Accounting, Emergy and Decision Making*, John Wiley, New York, 1996.
55. For example, a satellite mapping of the oceans shows that 41 percent is strongly affected by multiple human-made drivers of change (or imbalance). (Halpern et al., "A Global Map of Human Impact on Marine Ecosystems," *Science*, Feb. 15, 2008.)
56. Jens Peter Ellekilde Bonde and Jorn Oleson, "Interpreting Trends in Fecundity over Time," editorial in *British Journal of Medicine*, Feb. 16, 2008, Vol. 336, pp. 339–440.
57. "Cradle to cradle" is becoming a generic phase made popular by *Cradle to Cradle*, William McDonough and Michael Braungart, North Point Press, New York, 2002.
58. The EPA performance measurement system is at http://epa.gov/performancetrack/program/ems.htm.
59. MBDC's newest DFE certification criteria (2007) are at http://www.mbdc.com/docs/Outline_CertificationV2_draft.pdf.
60. A quick overview of the Wal-Mart Packaging Scorecard has been at: http://www.walmart stores.com/FactsNews/NewsRoom/6039.aspx.
61. Michael J. English and William H. Baker, Jr., *Winning the Knowledge Transfer Race*, 2006, McGraw-Hill, New York, 2006.
62. John Shook, *Managing to Learn*, Productivity Press, New York, 2008.
63. The official source for Kepner-Tregoe is the company itself: http://www.kepner-tregoe.com. The Altshuller Institute is probably the original home for TRIZ (http://www.aitriz.org).

64. After-action reviews (AAR) originated with the U.S. Army. The SOL coaching model emphasizing dialogue behavior is from the Society for Organizational Learning (Peter Senge).

65. Charles H. Fine, *Clockspeed*, Perseus Books, New York, 1999.

66. Punctuated evolution is the best-known legacy of the late paleontologist Stephen J. Gould.

67. The Johari window is from Joseph Luft and Harry Ingham in 1955. The name comes from Joe and Harry, and it has morphed for use in many contexts ever since.

68. Douglas K. Smith, *On Value and Values*, Prentice-Hall, New York, 2004.

69. Rick Reis, Executive Director of the Alliance for Innovative Manufacturing at Stanford University (http://ctl.stanford.edu/Tomprof/postings/464.html).

4

Dispelling Our Expansionary Habits

Compression will totally change the world of business, which at present can't cope with it. Some of its most basic beliefs are increasingly irrelevant, but still held as unquestioningly as if they were a tribal religion. At a minimum, expect screams of denial.

Why is such a change necessary and imminent? Business-as-usual does not equip us to meet twenty-first-century challenges, making even the twentieth-century split between socialism and capitalism seem passé. At the heart of that fracture were opposing concepts of property and ownership, and of the prerogatives and obligations that these conferred. Fairness arguments will continue because it is impossible to devise tax systems or funding systems that are fair to everyone. We do well to ameliorate gross inequities. But dramatically increasing our ability to deal with Compression is much more important than all that.

The contrasts in Figure 4.1 are stark, perhaps too stark. No real company's values are restricted purely to financial gain, and in a transactional world, no working organization can function without cash. The new organization is labeled a vigorous organization, defined as a learning organization that also takes action—it does things. Its ownership—how it is funded—is less important than its competence in learning and performing.

COMPRESSED IN THE COSMOS

The Toyota Way is a big advance toward a practical learning organization, but insufficient. Toyota's True North is "zero unsatisfied customers." Compression's True North is mutual survival with quality of life for

133

Dominant Values	Financially Controlled	Vigorous Organization
Concept of a company:	Financial assets	People: a set of stakeholders
Primary context of work:	Financial economy	Physical world
Primary purpose:	Make money	Carry out a social mission
Primary stimulus:	Competition; "winning"	Collaboration; recognition
Primary mechanisms:	Self-interest; transactions	Mutual interest; relationships
Primary measurements:	Monetary	Nonmonetary: space, energy usage, learning rates, etc.
Primary objectives:	Profit; return to capital	Long-term survival and quality of life for all stakeholders
Primary beneficiaries:	Ownership	All stakeholders; all humanity
Governance by:	Ownership board	Stakeholder board
Primary attention on:	Financial results; control for profitability	Human development; process integrity and improvement
Typical performance measures:	Growth in market share; asset size, profits	Quality of long-term outcomes for all stakeholders
Concept of operations:	Optimize profit	Process-learning laboratory
Organization structure:	Hierarchy and control systems	Flat hierarchy: self-operating; fluid self-adaptation to change

FIGURE 4.1
Contrast financial values and vigorous organization values.

all. This is incompatible with expansionist business thinking. But before going on, once again, what is Compression?

Assure survival of life and promote quality of life using processes that work to perfection with self-correcting, self-learning systems. No use of excess resources. No wasted energy. No toxic releases. Quality over quantity, always.

To grasp how different this is, imagine the life of 10,000 to 20,000 people in a radically different environment, aboard a large spaceship at least 2000 meters in diameter, isolated in the deep cosmos.[1] These descendents from normal earthlings must survive, stay sane, and find purpose in a harsh emptiness, light years distant from Earth's solar cycles and benevolent sun. Although our situation is far from this extreme, this scenario is instructive.

This space colony's economy could not replicate that on Earth. To be self-sufficient in deep space, this economy would obviously depend on the capture and control of energy. Energy captured from stellar radiation—or a preferred sun, if inhabitants had a choice—would power the recycling and reuse of everything, plus the replication of all the biological life cycles necessary to human existence. Exposure to either too much or too little energy would be disastrous. Unless able to generate matter from energy or to mine it from asteroids and planets, even feces and urine would be too valuable to jettison away. To build another spaceship, they would have to cannibalize material from the mother colony.

If truly self-sufficient, this spaceship would be sovereign, not a colony dependent on Earth. Although very high tech, work would be inseparable from the rest of life, like remote Earth-bound villages, or the life of a sea crew on a long voyage.

To preserve carbon-based life in an isolated container deep in space, imagine capturing energy and distributing it through all the molecular-scale gradients necessary to sustain life, maintaining a livable balance of oxygen, nutrients, water, gravitational pull, and other detailed necessities. A master brain (like HAL 9000 in space odyssey) can't centrally control anything this complex. Like a human brain scaled up too big, dense sensory input and "synaptic iteration" traffic would choke it.[2] This system would have to be decentralized, interactive, and largely self-regulating. Transactional market systems are simple by comparison.

Imagine a sci-fi scenario in which the energy level to sustain a population of 20,000 suddenly dropped to sustain only 10,000. Would inhabitants of the sovereignty collaborate to fix the problem, or degenerate into civil war, neglecting the systems on which all depend? If they collaborated, what reasoning would persuade them to do it?

Catering to expansionary mind-sets, many sci-fi scenarios conjure enormous energy, captured somehow, powering engines and weaponry with incredible propulsion. Energy shortages are a psychological downer. If serious, they spell physical doom. But excessive energy, or the wrong kind, is also dangerous. Repeated radiation exposure could hopelessly mutate all life within the sovereignty.[3]

Conflicts with space aliens akin to tribal conflicts on earth are favorite sci-fi fare. But deep in space, both dangerous aliens and vulnerable victims are at least a few light years away. Energy-intensive conflict would drain resources. The sovereignty's constant core concern would be to do or die on its own. Relegating a subclass of inhabitants to slave status to do grub work is not unthinkable, but developing a vigorous learning work culture would be crucial for sustaining life in this complexity.

Few basic needs to survive in deep space are actually known at present, and we may be able to foresee only a fraction of them. Earth's gravity could be approximated by centrifugal force on the outer wall of a huge cylinder. Humans would live mostly in an outer ring. A cylinder 2000 meters in diameter rotating at one revolution per minute would do this. However, air and "other stuff" would also press outward from the center. Moving toward the center axis a person would lose weight and enter thinner "air."[4] How this might complicate the sovereignty's internal ecology is unknown.

Humans can adapt to small differences from earth, like the Coriolis effect in a rotating cylinder (a ball thrown straight up won't appear to come straight down).[5] But can all forms of life, like those grown for food, exist without solar cycles and pollination cycles? Rotation speed might disconcert humans, too. Can we stay sane if our only outside view is looking down to see stars zip by in the dark?

"What gravitational level would you like this evening, my dear?"
"Evening? How many cylinder rotations from now is that?"[6]

Because a space-constrained craft could not nourish unlimited physical growth, humans could not spawn unlimited numbers of children, and nothing else could grow like kudzu either. Cramped conditions would stress both quality of life and quality of work. Life preparation would have to groom successions of people capable of the primary work roles necessary to sustain the colony. The young and the very old would drain the energy budget, but contribute little. Even if energy wasn't limited, thermodynamic economics would govern life and work just keeping the life-support systems in balance.

Economic trading would be trivial compared with constant vigilance managing Controlled Ecological Life Support Systems, the sovereignty's common "property." Priority number one would be sustaining life, and then improving its quality if possible. If distant from an energy source, milking every erg of available energy would be essential. If close, protection from excess energy would be equally taxing, and the sovereignty might have to cope with both excess and shortage at once. Permanent residence near an energy source similar in intensity and spectrum to the sun might make this easier. Without energy-garnering ingenuity, a space sovereignty could not maintain the basal metabolism to survive. An intriguing question is, if it acquired free energy to burn, how would it choose to squander it?

The space sovereignty's bioenvironment would have to nourish people well to develop their capabilities to the max. And in such a setting, what would private property consist of: some tools, personal care items, and a share of quarters? Would *ownership* mean primary responsibility for operations in a corner of the craft, with an obligation to pass them on in better shape than they were found? With life itself riding on competence, status would accrue more because of what people *do* and help others do, than because of how much they *own*.

Of course, this is little more than sci-fi speculation as a somber illustration. An "Island Three" sovereignty in deep space is far beyond present technology. Our most venturesome scenarios imagine solar system colonies impossible without big breakthroughs. We have little actual experience on which to base them. Besides junk in near-Earth space, NASA notes four major risks in prolonged space travel within the solar system, as with the manned Mars mission: (1) loss of bone density in a weightless environment[7]; (2) solar flares and other cosmic radiation endangering all life on board, not just human[8]; (3) small particles in space penetrating the skin of the spaceship[9]; and (4) behavioral and psychological problems of teams functioning under stress for long periods in a confined environment. Any of these risks, plus those unknown, could doom the NASA Mars mission.

The fourth known hazard has a high probability. Teams sequestered for long periods on Earth, as in remote labs, sometimes experience severe psychological stress. Tight confinement in abnormal conditions in a spacecraft would likely exacerbate this.

For example, eight volunteers confined inside the Biosphere 2 experiment dome for two years on *Earth* incurred significant problems that split them into two camps.[10] Can-do astronauts can tough out behavioral stress for a few months, but those heroics disguise the fragility of astronauts' psychological health, despite careful screening. Living from birth to death in remote space is literally another world of challenges.

Most knowledge of human sex in space is anecdotal; for example, that females on space teams calm team behavior. Earthbound couples such as in Antarctic research lab teams are known to have paired off for sexual liaisons to relieve frustration, ceasing at mission's end. The strange case of Lisa Nowak allegedly attempting to murder a romantic rival blew the lid on sexual attractions among near-Earth space crews.[11]

Reproduction is crucial for permanent life in space, not only for humans, but for anything grown for food or for other needs; but little is known about biology in space, much less raising children there. To assume full responsibility, each generation of children would almost need to grow up learning the work. Escaping our biological dependency on earth is so far beyond our technology and human capacity today that to adapt, all life might have to evolve differently. Some seriously think that we need to enhance our human capabilities to progress on Earth, too.[12]

We are already on the High Frontier, Earth, a robust space sovereignty circling the sun. The High Frontier is us—our systems for work and our means of governance—or nongovernance—of the planet. Thinking of Earth as a space sovereignty shifts our view of what we must do and what we must become within our own ecosphere. However, most private space ventures today are still inspired by expansionary business tradition, in the hope of making a buck. The Space Frontier Foundation promotes and tracks many of them, mostly inexpensive satellite launches and proposals for high-ticket space tourism thus far. Severe energy shortages bode ill for energy-intensive manned space flight by present means, but grand dreams don't die: mining riches in the solar system, and manufacturing in a weightless environment (perfect crystalline lattices, for instance).[13]

However, by viewing Earth as a space sovereignty with limited resources and using thermodynamic logic (long-term energy return from energy expended), space ventures will be more realistic. What can we learn or obtain from the energy spent? How much can we miniaturize loads lifted into space? Space entrepreneurs complaining about regulation are actually signaling that they face a much more complex situation than aviation pioneers a century ago. If they refuse to think about this complexity, regulators are forced to divert their single-track thinking.

To cope with Compression we need to radically change our thinking about how work is organized and the goals of work organizations. Executives struggling with financial woes may not think so, but expansion has been a mistake-tolerant utopia fueled by cheap energy. Compression is a dystopia. It calls for discipline organizing human skill and learning that is nearly impossible to sustain within the present system. Forget easy-street idealism. Space sovereignty could not sustain itself using today's economic system. We can only learn by making mistakes, but hopefully we can learn to make smaller ones, and catch them earlier. No precise formula for this can be proposed, only a general direction. Following its own learning methodology, such a system has to evolve rather than be constructed, beginning with the work organizations most crucial to quality of life.

THE BLINDING MIND-SET

The first of many blind spots in transactional economics is assuming that transactional valuation models are sufficient for guidance. Of course, any system used is based on human values, but dollars, euros, and yen are all standards of human value for transactional exchanges and the arithmetic by which we assign human value to assets. However, as measures of human preference, they are inadequate for managing physical processes important to everyone's survival as well as welfare—the situation in Compression. That's due to the mind-sets that transactions create and to the work organizations created by default using transactional guidance.

Transactional mind-sets blind us to the obvious. For example, ask children to describe a home loan closing, a transaction too complicated for them to comprehend. They describe what they see: people talking, shuffling papers, signing—physical activities. Adults concentrating on the transaction, oblivious to the obvious, barely notice this. Once our engrams to analyze transactions are activated, the brain's executive functions screen out much physical activity associated with them.

Transactions are the fabric of industrial society life, but being only markers of human value, they blind us to the processes that created the results we exchange. For example when we eat at a restaurant, we evaluate our experience, whether food quality and service were worth what we paid. All the processes to source and prepare the food may be totally unseen. Stock trading is even more abstract, an exchange of valuations of expected future income. Tradings of financial derivatives are exchanges of representations of future income that are many stages of abstraction removed from any original basis for their value.

Mentally, we objectify transactions. That is, encasing vague concepts of processes in mental wrappers converts them to items to buy and sell. We encode a personal service like hairstyling into a mental package for a price. Parties clash if their concepts of a contract's terms differ, so people will quibble over commas in complex contracts like those in house building and insurance. However, few of us digest the fine print of routine approvals—as when clicking off the conditions to use a wireless network. Once inured to objectifying items for exchange, we can accept the abstract logic of intellectual property creep, like copyrighting the hairdo of a beauty queen.

Once upon a time, transactions were many fewer and much simpler. A nineteenth-century rural household might shop once a week when they went to town. Occasionally peddlers stopped by. Most houses were built by local labor using material at hand. Growing food, storing it, preparing it, and eating it at home had its problems, but transactional complexity wasn't one of them. A modern suburban family easily logs more transactions in a day than a nineteenth-century farm family in a month. An online stock trader may do it in a few minutes. To save human time, transactions had to automate, from E-ZPass toll roads to automatic bill payments. To speed transactions, auto-ID of people with natural biomarkers or embedded radio-frequency ID chips raises the specter of Orwellian control.[14]

Ancient merchants expected buyers to haggle, to have a social exchange. Hometown shopping was a social occasion. Now we lack time for that, whizzing about, the vapors of our transactional contrails lingering as evidence of where we were and what we did.

But monetary valuations and transactions activate physical processes far removed that in turn affect many natural ones. Many of these linkages lie outside the perceptual capacity of transactional man, so distinguishing business processes from physical ones is useful.

- Business processes: negotiations, valuations, and transactions
- Physical processes: both human controlled and wholly natural

During expansion, our transactional fog kept thickening, obscuring what we actually do, but sometimes we broke through it. One example was Toyota's developing the original TPS in the mid-twentieth century by changing what they physically did, minimally inhibited by business conventions. A business plan reduced to pro forma dollar projections only helps to assure that a working organization will make money. It need not imply a social mission for the organization, nor how it will learn to carry out that mission better and better using fewer and fewer resources.

DYSFUNCTIONAL ASSUMPTIONS

All the assumptions in Figure 4.2 are stereotypical. Exceptions abound. Many socially responsible companies have codes of ethics and tests of performance other than dollars. Many use a balanced scorecard (returns to owners, customer satisfaction, internal processes, learning and growth).[15] Environmentally conscious companies have the triple bottom line (planet, people, profit). Others stress service to customers by criteria like those of the Baldrige National Quality Award. Innovation awards such as PACE (for innovation by auto suppliers) recognize significant innovation, defined as "game changing."[16] ResponsibleShopper (http://www.responsibleshopper.org) rates social and environmental activities, and the Great Place to Work Institute rates places to work. *Fortune* even has a blue-ribbon list of companies on lots of other lists.

However, executives with tunnel vision on profit regard such lists as beauty prizes, and no organization in a transactional economy, even a church or a school, can ignore its finances. All die if they run out of cash. Somebody must supply every work organization's cash kitty. Any organization in debt to a normal financial institution is tied by loan conditions to global financial markets' demand for return.

Finance is the business world's control system. That's why transactional thinking dominates process thinking in that world. Process thinking concentrates on human and physical process relationships. Transactional thinking concentrates on results, crowding out time to think deeply about people and process interrelationships.

Few managers are as wedded to business thinking as implied by Figure 4.2, but without being resisted, this thinking presses them all to favor quantity over quality. Today's industrial economy functions by markets and transactions. Thoughtful business leaders know that decisions based only on financial calculus can be destructive, but they are tethered to financial rituals. Consequently lean operations are viewed as only cost cutting. Green initiatives must assure managers that environmental sustainability will not bankrupt the company; financial peril seems more real than physical peril. In addition, budgets fragment people in different departments, and profit-and-loss boundaries separate people in different companies that should be

Assumption	Brief Description
Unlimited growth	Nothing except market acceptance limits growth; resources have no long-term limit. (Abetted by linear financial models.)
Economy of scale	Bigger is better; revenue will be higher and fixed cost will be spread over more units.
Short-term bias	Money sooner is worth more than money later. For anything not coming to fruition for fifty years or more, payoff is so near zero why think about it.
Results bias	Concentrate on sales, costs, and representations of wealth, not on how these results come about or consequences of them.
Specialization (inflexibility)	Doing separate operations well at lowest costs adds up to the lowest total cost. Worry about set-ups, changeovers, and adaptation only if necessary.
Fragmentation and tribal	Divide work by stations or departments and cost-control each.
Complacency and panic	If profits are high, everything must be working well; relax. If losing money, take emergency measures to resume profitability.
Commodity traps	Grow markets by low prices, competing on cost (Walmart model). If companies wind up with a high break-even point, a small sales dip sinks profitability.
Market rationality	Buyers and sellers both understand what is being sold, and that profitably meeting a market test is proof of effectiveness, or even efficiency.
Model fixation	A hyperrational model (like cost and financial statements) is the reality it purports to represent.
Ownership dominance	The main purpose of a work organization is to make a return on ownership capital, so ownership should have ultimate control, with managers as agents controlling operations in the owner's interest.
Money as the common language	Money may not guide every decision, but most activity has to be reviewed for impact on the bottom line.

FIGURE 4.2
Assumptions hidden in expansionary business thinking.

closely collaborating to integrate complex products and systems. Leaders sense that much better is needed, but can't escape the clutches of money-think.

Unlimited Growth

Ask almost any business person the objective of the business. The response is apt to be either to make more money or to grow the business, assuming that any business not growing is stagnating. Exactly what a business leader envisions by growth may be murky; it may be top- and bottom-line financial growth, or it may mean new technology, service, products, or even a new business model without much change in financial size.

Representing limits in growth of natural resources is almost impossible in linear financial statements or spreadsheets. Simple financial projections suggest process scale-up in the same proportion, so strategists rely on nonlinear logic to encode changes in processes or business models into financial pro formas. It is natural to desire growth; it makes managers' problems happier than if struggling to stay financially solvent. But both of these conditions are very different from leadership to create the most proficient, fastest-learning organization possible.

Economy of Scale

Assumptions of unlimited proportional growth are hidden in the simplest of business formulas. Take that old standby, the break-even point formula:

$$\text{Break-even volume: } V = F/[P - C], \text{ where } F = \text{fixed cost,}$$
$$P = \text{price}, C = \text{variable cost, and } [P - C] \text{ is margin}$$

Similar logic estimates the time until an investment is recovered, after which all profit pays a return above the investment—what is really wanted. Longer times to payback are considered riskier; more things can go wrong. Payback sooner and higher beats later and lower.

This simple, straight-line break-even formula abets expansionary bias, short-term planning horizons, and sometimes the illusion that a business can scale up indefinitely. Smart business strategists don't fall for these. However, they may not foresee its emotional traps: euphoria if profit margins are high (low break-even point), and complacency if margins are low with high volume and a high break-even point (commodity trap).

Straight-line graphs also suggest that maximizing sales from an investment (fixed cost) will maximize profit. Curving the equations depicts pumping up sales volume by cutting the margin (until marginal cost equals marginal revenue). By this logic as margin nears zero, break-even approaches infinity; therefore, grow to the limit of capacity. But then as noted earlier, if volume turns down, a high break-even point with puny margins sets up a crash. To stave off collapse, costs once regarded as fixed become variable, subject to chopping until revenue once again covers costs. If the chops are also linear (cut each expense by 20 percent), impairment of operations is unpredictable because all expenses are never equally important to the whole.

The overall fallacy here is regarding a company as a financial machine, not as a work system or as organized human capability, dependent on human performance for communication and learning—and payback from the time invested in learning is difficult to project financially. In addition, the effects of quality, flexibility, innovation, and tight integration are poorly captured by a linear financial model.

Short-Term Bias

Almost all executives bemoan quarterly earnings advisories pressing them to show short-term monetary results. The flip side is disincentive to invest in longer-term objectives. Consequently, accounting games to smooth earnings or defer losses have been standard practice, creating constant tension over acceptable accounting practice even when no one intends to deceive.

But the roots of short-term bias run deeper than Wall Street expectations. For instance, take that staple of transactional calculus, interest. In industrial societies, every literate adult is somewhat familiar with simple interest rates, and perhaps compound interest. Compounding pays interest each year on the original amount plus on all prior interest accumulated to that time. Adults usually know the basic version of compound interest rate growth: $FV = PV \times [1 + i]^n$, where i = annual interest rate, FV = future value, PV = present value, and n = number of years.

School-trained managers also use the inverse of this, the present value of future payments: $PV = FV / [1 + i]^n$.

The magic of compound interest awes those grasping the growth curve of this model for the first time. Small investments swell into huge sums ($1 at 7 percent for 10 years is $1.97; at 100 years, it's a whopping $867.72). Ancient philosophers and prophets condemned interest as usury because at ancient rates of 30 percent, *simple* interest balloons a $1 loan into $4 of debt in only 10 years, and nearly $14 if compounded. That's hopeless to pay off with ordinary earnings, so capitalist expansion depended on rates that were possible to repay.

During expansion, longer-term development loans had interest rates closer to 2 to 3 percent, while 6 to 10 percent compounded became the long-term return in stocks. Playing the system, borrowing money at 6 percent and investing it in stock at an 8 to 10 percent average return, generates monetary wealth through financial leverage. Plenty of people have astutely leveraged borrowed money by investing it in high-risk, high-return investments. This works as long as the physical economy can expand to underpin the financial swirl that supposedly represents it. However, as can be seen by recent financial collapses, leverage in reverse can bulldoze paper wealth very quickly.

Besides the unpredictability of market volatility amplified by leverage, straight present value calculations make the value of any dollar received more than 10 years out seem trifling. It makes long-term financial planning seem to be a waste of brain power. Despite this, leaders do try to develop people to assume responsibility decades hence, and some of them plant trees. When they do, they are not thinking financially but preparing to meet challenges impossible to specifically forecast. Preparation for unknown challenges is beyond financial short-termism.

Results Bias

Accounting models distort perception. Because anything not transacted is not in the model, use of "free" natural resources is not captured. And lean conversions battle accrual accounting that does not monetize its benefits. For instance, if cutting output to shrink inventories does not absorb the planned overhead, the variance may show a loss even though the cash position improves. These kinds of distortions have provoked a lean accounting movement.[17]

Transactional thinking multiplies everything by a cost or price, painting it the same color with a currency brush, and mentally converting everything into objects for trading. Besides the bias of human valuation, this objectification tends to mask interrelationships among the variables thus quantified: out of model, out of mind. The system multipliers create additive quantities but cannot detect process relationships. For example:

$\Sigma[\text{Units} \times \text{Price}] = \text{Sales \$}$
$\Sigma[\text{Units} \times \text{Cost}] = \text{Material \$}$
$\Sigma[\text{Units} \times \text{Cost}] = \text{Labor \$}$
$\Sigma[\text{Units} \times \text{Cost}] = \text{Energy \$}$
$\Sigma[\text{Units} \times 0] = 0 = \text{Effluent cost or value, nothing}$

In addition, managers have a predilection for using cost percentages and cost ratios. For example, because of the dollar multipliers, a units/person ratio is not the same as a cost productivity ratio, the inverse of which is a cost percentage, and presumably the lower the better. That is the beginning of distortion, however. Labor rates vary depending on how big an overhead percentage is piled on them, and people who should know better can deceive themselves that eliminating an hour of labor cuts cash expenditures more than it actually does, a simple, well-berated error. However, as we try to interpret more and more interactive complexity from these additive models, the weaknesses become glaring. They concentrate attention on financial results, not on what we really need to do.

A great deal of effort has been spent devising "more accurate" cost accounting systems.[18] Some are less distorted than others, but less bad is not really good. Effort is going into devising annual reports that reveal more than matters important to investors, too. But the core of the issue is results-oriented myopia so narrowly focused that people can't relate to anything without a dollar sign in front of it.

We can do better. Process factors and relationships can be expressed, if crudely, in nonmonetary form; for example, energy yield, resource yield, or recycle percentages. As long as we have some form of dollar democracy, with markets giving customers preferences, we'll need measures useful for transactions, but in Compression we have to see outside the boxed-in models that drove expansion.

Results-oriented management usually demands financial results described by some standard, acceptable financial accounting system. But in the complex world of Compression, this is dysfunctional. It's a bit like a two-year-old squalling for a momentary whim, totally unaware of the processes necessary to fulfill them, or of any undesirable consequences if he actually got what he wanted.

Specialization and Inflexibility

Cumulative growth of a self-similar expanding organization eventually limits its capability. The structure that developed to execute an old business model is stressed as it takes on activities for which it is less suited. Size slows its ability to respond quickly. Mission creep adds complexity to the work. Nostalgia drag slows its ability to change into something else. ("This is not who we are," "Customers want us to keep doing what we do well," "We have always done it this way," and so on.)

Regarding an organization as a money machine tends to calcify it by setting it up to assure that money flow is going according to a financial plan, or budget. Once cranked up, controllers are reluctant to risk slowing a money machine with changes that threaten existing cash flow. New directions that could cannibalize existing income are resisted. The classic mechanical analogy is an expensive automated machine designed to cut only one size and type of engine block, useless for doing anything else. Once built, this monster must be fed, so the company must desperately try to sell the volume that keeps it fed. That's quantity over quality, and certainly not flexibility.

This mechanical analogy fits almost any financial targeting. Financial plans range from earnings advisories to departmental budgets. In any case, meeting the plan tends to have high priority. If performance falls short, was performance poor, planning poor, or both? In any case, unwavering demand to execute a financial plan impedes flexibility to deviate. Given the kinds of challenges coming in Compression, the need is for fast learning and flexible response, not thinking of a work organization as a money machine.

And why do we need flexibility to take on doing much more with much less? For example, how would one—

Convert millions of heavy SUVs to fuel-sipping lightweights?
Move as many people through an air traffic system using less fuel and less waste and with limits on available landing space?
Prepare corporate and urban systems to survive natural disasters—like an Asian tsunami or Hurricane Katrina in New Orleans?

Retrofitting old thinking organizations for new challenges has not been very effective. A typical expansionist organization's life-death cycle is birth, growth, maturity when complexity adds waste, then fade-outs when markets or technology passes it by. That splinter-and-reform model served us well, but now we need working organizations that learn more rapidly for faster transitions with less turbulence.

Fragmentation and Tribalism

The difficulty of communicating between functional silos is now a well-worn cliché in management discussion, but too little has been done to improve it. The same issue clouds communication between all kinds of work organizations. Intellectual property tollgates limit the density of communication between customer and supplier companies, too. For example, purchasing agents negotiating big-volume contracts may pretend that an auto company just buys parts or assemblies from suppliers, and then press them on price. But someone in these companies must closely communicate to integrate the design of a vehicle in which all components must complement each other, rather than a collection of parts hung on a common body. Here again we can do much better. For instance, Chrysler once formed suppliers into a virtual design company to develop the instrument panel for a new vehicle. After all participants took off their company hats and donned the virtual company one, this worked quite well, but so far as is known, this exercise was never repeated.

Organizational isolation strengthens the affinity of people to local groups or to work specialties, rather than to a common mission. Transactional boundaries and intellectual property exacerbate possessiveness between companies. Budgetary separation exacerbates it between internal departments (functional silos). If intense, these rivalries resemble warring tribes, each defending its turf and sometimes invading that of others. In addition, if each silo is run by a management pyramid, human energy is wasted competing for alpha status. When fragmented working organizations are formally bonded by no more than fair transaction rules for each to independently make money, collaboration is inhibited. Given all this, the degree to which they now collaborate is remarkable. Figure 4.3 may illustrate this effect better than words.

Inability to Simplify Complexity

As an example, automotive drive trains able to simultaneously achieve high mileage and near-zero emissions must blend all parts so that each

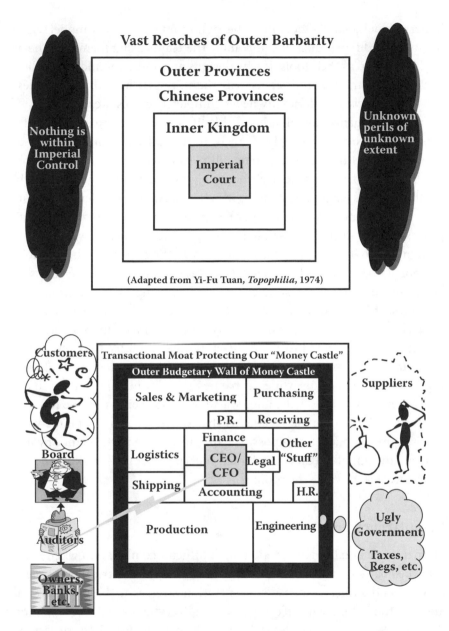

FIGURE 4.3
Comparing an ancient Chinese mind map of the world, 500 BCE, with a stereotypical financially-driven company.

one reinforces the roles of every other part (higher free-energy density designs). To do this, development processes, computer simulations, and hyperrational analysis tools must be tightly integrated. All models used must be valid representations of reality. The teams using them must be as well integrated as their tools. If new designs are cradle-to-cradle, new life-cycle business models are designed with them. This is a huge shift from sales-are-final models. Responsibility never completely ceases. Responsibility of every work organization all around this life-cycle chain exceeds that of a supply chain today. (A supply chain is all the linkages from suppliers to final customer.) In a life-cycle chain, all players have to perform superbly in all phases of their work. Delivering a good part is not good enough. Neither is arm's length negotiation, which is presumed to offer fair opportunity to all comers. Any newcomer has to demonstrate ability to integrate processes. Low-cost bids to sell items cease to mean much.

Complex contract clauses cannot anticipate the role of supplier flexibility, nor its benefit. Neither does taking bids, done to assure fairness in awarding contracts. For example, a low bid winning a complex construction project is just an entry point. Discovering the problems converting an architect's drawing to reality, seldom easy, is made more complex if owners, architects, and historical societies make multiple changes during construction. In many cases the real service is deftly executing engineering changes on the run while all parties thrash out what they actually want or need. Obviously, the huge waste in these projects can be reduced by improved learning; but as long as that waste is generating large variances, a precise budget is impossible, and measuring performance by it is not very meaningful. (Charging for engineering changes—making money on the waste—is where the profit is.)

Complacency and Panic

Human logic and human emotion are a difficult combination when work must be nearly perfect, like that of air traffic control, electrical power generation, and surgery. Intense concentration stresses us beyond physical and emotional limits; it can't be sustained. Everybody needs recovery and diversion time. Work in systems that must be near-perfect all the time must compensate for this human fallibility. Industrial society now has many such systems, and their number is unlikely to diminish in Compression.

Organizations that must both innovate and operate near-perfect systems crucial to our welfare have to concentrate on performance. The emotions of boom-bust financial performance easily distract people from this.

Spectacular business failures from Long-Term Capital Management to the subprime liquidity crisis feature overconfidence followed by collapse. The urge to win warped judgment. Financial success bred complacency, followed by bigger and bigger gambles trying to keep the game going. In the Long-Term Capital case, traders ventured further and further outside the risk assumptions of the company's trading models until they blew up in their face. Similarly, in the subprime mortgage mess, traders came to believe that global capital markets could absorb any risk package they dumped on it.[19] When euphoria of winning turns to panic to avoid loss, people often stop learning and do as they always did, like compulsive gamblers assuming that eventually the tide will turn. In the Enron case, managers pressured for financial growth conjured schemes of financial fantasy.

If managements believe that profits are a reward for being ahead of the pack, when ahead they relax, enjoy life, maybe even present their case to an MBA class. But when performance must be sustained at near-perfection and learning is essential to keep doing it, such complacency is disastrous. Human temptations to break concentration have spiced the plot of many a movie, as with champion athletes. Paying too much attention to money, or being compelled to, is one of those distractions. The psychology of meeting the challenges of Compression is the opposite of striking it rich.

Commodity Traps

A high-profitability business (low break-even point) may become complacent in grand style. One with a high break-even point and low margins must sustain high volume to keep its size, and maybe even to keep its core capability intact. It is in the commodity trap if it lacks the resilience to shrink while sustaining excellent performance.

When an entire industry is in a commodity trap, competitors struggle to maintain physical property and core organizational cadre. Instinct is to strip every cost not obviously required to keep running—at the highest volume possible—trying to drive each other to the brink, all descending into mediocrity—classic quantity over quality. Major airlines and the auto industry appear to be in this trap today. Cost cutting is from the view of ownership: preserve as much capital as possible as long as possible. Despite protestations to the contrary, organizational performance diminishes: provide less service, cut training, defer nonessential maintenance ("if it ain't broke, don't fix it"), outsource operations to less expensive contractors, shed excess assets, and so on. And of course, if rigid work rules had become monuments that decrease flexibility, workers instinctively want to save those rules, thus increasing management's incentive to outsource around the monuments.

In Compression we have to confront this differently. Expansionary bias says that slow growth or shrinkage is only temporary. Suppose it isn't; suppose that learning to do much more with much less is becoming the new normal. Then many organizations must decide what they do that is critically important to improve—their social mission—and seek to perform it very, very well. Because money is necessary for this, leadership has to both find the money and cut the fat. That's the situation of hospitals, schools, airlines, utilities, railroads, and maybe manufacturers of light bulbs. Both investors and the organization have to support performing the mission with excellence rather than dream of future bonanzas. That's a profound psychological change.

Of course, organizations in a commodity trap can cut costs by simplifying processes, simplifying product designs, reducing error rates, and so on. But once engineered lean has captured the easiest gains, the next phase depends on developing a core of able, dedicated people. Long term, the challenges of Compression will take astute, organizational problem solving the likes of which have not been seen before. The real challenge to business thinking is to create business models that can cope with Compression in a transactional economy without hanging up in commodity traps.

Market Rationality

Market rationality is based on shaky premises: that millions of people voting with their money represent the wisdom of a crowd, and that all buyers and sellers have equal information when evidence to the contrary is so overwhelming that it's the main reason for regulating markets. We prohibit insider stock trading so that people in the know can't take advantage of less privileged traders, all of whom are presumed to be equally ignorant. Investment letters have readers, and financial counseling is a career. For consumer purchases, many people trust *Consumer Reports* or Angie's List over advertising claims, the Yellow Pages, and tips from friends.

Recent history suggests that expansionist market speculation still distorts simple price signals. Speculation on every risk angle of both spot and futures markets fuels boom/bust volatility in petroleum, gas, and metals market. Other examples are 2001 faux markets for electricity in California, carbon credit trading in Europe, and world food prices. In the latter instance, index fund holdings of wheat futures contracts had risen to 40 percent by the end of 2007, suggesting speculative overdosing of a formerly staid market.[20] Nongovernmental organizations and governments started bypassing markets to form agreements among themselves to alleviate severe food shortages. Volatility decreased, but farmers who need to tend other concerns are still forced to gamble to decide how to sell grain. This market system self-creates a lot of imaginary complexity.

In Compression, expansionary markets will become increasingly dysfunctional. If the supply of fuel, fertilizer, electricity, grains, fish, or anything else cannot be increased at any price, classic markets for them no longer exist. Then a market funnels what supply there is to those with money, and denies any supply to those without it. One can argue up and down about self-sufficiency, taking personal responsibility, and so on, but when a system starts literally starving people to death, they won't tolerate it.

The only fair solution is some form of rationing to cap aggregate consumption. Equal allocation per capita seems fair; administering it fairly is more complicated. Suppose every licensed driver received coupons for, say, 10 gallons of gasoline per week, but could sell them to others. Could this work using credit cards? Commercial users would need a higher allocation; counterfeiting and other cheating would have to be countered; and so on. If drivers switched to alternate fuels, those might have to be rationed, too. Everyone will not regard any system as fair, just as they don't regard any tax system as fair. Loophole-seekers will lobby in full force.

Cap-and-trade emissions markets, a version of ration trading, foretell issues to be expected. Expansion-minded marketers enthusiastically balloon the markets, forgetting the objective of reducing aggregate fuel consumption.[21] Dealing with Compression takes more than a mechanism; it takes a different mind-set.

Indeed, people will not accept rationing unless they understand why it is necessary. But if convinced that everyone's welfare depends on sharing a common resource, most people will pull together, just as most Americans put up with rationing during World War II. To accept permanent restrictions, people need to be confident that it is necessary, and that governance of them is not shot through with favoritism. At present, no institution—corporations, governments, or media—fully enjoys that level of public trust. Social missions have to take priority over money making before they do.

Many plans will be proposed. For example, Herman Daly proposed indirectly capping all consumption by tying money supply to land trusts. Land is fixed in quantity, and a trust is a contract to exercise stewardship caring for some part of it. Executing this might hit snags, too. If this currency did not affect specific market behavior, it might not limit consumption of specific foods, fuels, or other items—or prevent skewing income and consumption to the rich.[22]

The simplest human systems to cope with limited supplies are the social codes of usufruct mentioned in Chapter 1 as applied to water. These codes have lasted for centuries, but all the societies in which they prevailed operated by variants of feudalism. We need a vision that improves on that if we are to extend an industrial society quality of life to all.

We have to have a different mind-set to do something very different. Expansionary bias appears to be degenerating free enterprise into a caricature of itself with everything regarded as a market. Both megachurch services and black social activism mimic the entertainment business, and not even the reclusive Amish totally escape it.[23]

Kicking market hype is like trying to stay drug-free in a house full of pushers, but feeble movements to do this have begun. Some sell ad-zapping technology, assailing the system directly. Others encourage swearing off commercial opiates cold turkey.[24]

Model Fixation

Learning at work today has to include validating abstract models with reality. Model coolness does not signify reality, as discussed in Chapter 3. Most of us realize that the seemingly real worlds of computer gaming, media entertainment, or fantasy sports are unreal even if we are addicted to them. But for example, it is not so easy for an architect to detect that his computer model is designing a castle in the air unrelated to the realities of either building it or of operating it. He or she needs experience outside these models in order to question them.

Sometimes almost all we know is from hyperrational models and images. All we can do is constantly question whether they are consistent with other information, much of which may also be obtained hyperrationally. That is, does all the information coming into us paint a consistent picture? If not, what suspects do we check out, and against what standard? More than consistency of data or other evidence must be considered; do assumptions in the models conflict?

Indeed, macroeconomics consists mostly of studying aggregate fragmented market behavior. Although critiques of GNP models are legion, it is almost always assumed that the bigger the aggregation, the better off everyone is—quantity over quality. At the microeconomic level, the test of a new idea is whether a financial projection of a business model for it will yield a return, allowing for risks. Rarely are the basic assumptions of this limited view of the world questioned.

Ownership Dominance

Nature forms hierarchies. Animal groups have pecking orders, and ship captains give orders, but micromanagers overloaded with information are incompetent. Effective leaders have always had to delegate responsibility, just like emperors in ancient Rome or Beijing. None could possibly control an empire in detail, so distant officials adapted a common imperial pattern of governance to changing local situations. Once developed the Roman way, distant praetors usually ran provinces that way to benefit Rome (although uppity ones also marched on Rome). Making a change in the pattern was a big imperial deal.

Large work organizations still work that way, with a chief executive as emperor checked by some governance system. But in the case of public-stock companies, are the capital markets really the impersonal emperor to distant CEOs and CFOs who are more like praetors running organizations primarily to benefit capital markets?

In small businesses, like proprietorships, the owner's final say is accepted; they personally assume the most risk. Anyone financing startup entrepreneurs, large or small, wants them to have skin in the game to ally them with the financiers.

However as companies grow, ownership becomes more diffuse. Faceless stockholders expect returns. Creditors impose covenants. Top managers better not miss many payments or let the stock price tank. High-flying stock generates paper capital to use purchasing other companies with stock swaps, for instance. If deal-making euphoria grips boards and managers, they can amass more organization than they can lead under one financial umbrella. If zapped by hostile takeover bids, they are legally obligated to maximize long-term return to stockholders, which often means selling out to the highest bidder.

This system assumes that people with money to invest are more likely to be responsible citizens exercising better judgment. It also assumes unending expansion; the prescription for economic downturns and most corporate ones is to resume growth. But as can be seen by recent financial bailouts of banks and other concerns, when an organization's functions are vital for all society, but it can't continue to fulfill them, society (government) has to intervene. This starts to clarify that such an organization's performance has more value to society than its returns to ownership. And if everyone has skin in the game, distinctions between public and private capital blur.

Everywhere such an organization operates, people depend on its jobs and money. If its services are crucial to our quality of life, performance of its mission affects everyone, perhaps everyone and everything on earth. If its leaders take this seriously, they not only have a public service obligation, but so does everyone working in the organization or contributing to it in some way. Financial viability is merely a necessity to fulfill this mission to *all* stakeholders.

Many workers have more relative financial livelihood at stake than distant investors. For example, workers' retirement and health plans based on employment likely depend on investors in distant capital markets. In an industrial society, high-tech systems do more than augment informal social compacts (kith and kin will care for you). Kith and kin may care about you, but they seldom interpret DNA profiles or perform colonoscopies.

Nobody wants mission-critical work organizations to fail, financially or otherwise. We want organizations like fire departments to perform to perfection when needed, but not to maximize profit. We don't want them to be a growth industry, nor do we want to be their emergency customer. But unless we experience fires, we may not sustain a fire organization in a peak state of readiness. We drift into our target-risk zone, taking our chances that with minimal care, fires won't happen to us.[25] We would rather buy things we prefer than support common resources—quantity-over-quality reasoning again.

In Compression, many work organizations—transportation, vehicles, electrical power generation, construction, even carpet companies—have to be regarded more as fire departments. For example, Interface, Inc., a carpet company, has for some time been working toward environmentally sustainable business operations. Carpet is designed to be attractive, totally recyclable, and safe.[26] To recycle, the business model needed a complete makeover to take carpet back, but Interface is still far from carbon neutral. The objective is "zero by 20," meaning by 2020 to use zero virgin raw material, zero energy not generated on site, and zero toxic releases. The mind-set to sustain this strategy through the financial fog of an expansionary, transactional economy is Interface's most difficult challenge. It is subject to the same impersonal demands for a rate of return as everyone else.

Marjory Kelly, for one, thinks that as long as the legal purpose of corporations is presumed to be maximum return to ownership, challenges like dealing with Compression will always be compromised. Both law and market expectation bias key decisions in favor of growing ownership wealth. She likens this bondage to the divine right of kings, a doctrine whose time should have passed.[27] However, faith in it is so great that many people favor privatizing government functions, presuming that the profit incentives of "private government" will better serve all stakeholders.[28] However, many conflicts between public benefit and private gain seem irreconcilable. As an extreme example, how can R.J. Reynolds simultaneously advise youth not to smoke and assure investors that they are growing a future market for their product?[29]

Corporate support of communities and charities does not compensate for profligate waste making money. For example, while sinking in their own self-deceptions, Enron executives were major benefactors of Houston. Even Pablo Escobar, ruthless capo of a drug cartel in Medellín, Colombia, was a Robin Hood giving generously to soccer fields and housing for the poor.[30]

Any new system has to break our gridlock on old arguments: labor versus management; social responsibility versus profit maximization; individual freedom versus social control; control for predictable financial performance versus development for the future. Whenever earnings disappoint, capital market whips still come out—deliver your prediction or your stock price takes a beating. General Electric, an earnings darling, rediscovered this at the end of the first quarter, 2008. One hiccup damaged GE's credibility with financial analysts.

Predictable growth is a financial tranquilizer with a side effect—volatility. Small swings in profit-to-expectation magnify swings in stock cap value, where the big money resides, but inside companies, profit-assuring control systems sap free energy. These contradictions run deep. I have twice observed shop-floor folk bemoan controls (and a sale of their company), but suddenly switch when seeking the best return for their own retirement fund. They saw no connection between this and their work experience.

From earning advisories to detailed budgets, financial controls create the illusion that hitting financial targets signifies success—but success doing what? What they really do is emphasize results over innovation and quantity over quality.

In recent years, business leaders have begun to sense that something is wrong. Financial reports for investors don't speak to the interests of all stakeholders. Accrual accounting distorts the financial effects of streamlining work processes. A movement called Beyond Budgeting strips away detailed budgetary control, enabling more decentralized management with much less command and control.[31] The movement toward corporate social responsibility is moving well beyond community support, philanthropy, and token environmentalism.[32] By 2008, world movers and shakers at Davos, Switzerland, were beginning to recognize that tokenism could not cope with resource shortages, environmental issues, and stakeholder unrest.[33] One of the more aggressive proponents of genuine corporate social responsibility is Corporation 20/20, sponsored by the Tellus Institute. It holds that corporations must cease absolute allegiance to ownership, so their strategy can shift from "the business case for social responsibility" to "the social purpose of business."[34]

As long as we are gripped by the idea that the main purpose of business is to make money for ownership, tinkering with financial formulas is useless. Companies regarding the challenges of Compression as a side issue will resist serious changes to meet its challenges.

In expansion, when resources were plentiful, profit maximization worked. Almost any system worked in a fashion, including the Communist ones, but requirements in Compression are more stringent. We need vigorous learning organizations.

Money as the Common Language

The common language of business is money—finance, accounting, and marketing. Aspiring employees get an MBA to learn it. They can't communicate with top management without it. In a transnational society, money is the common language of governments and nonprofits, too; doing almost anything takes money.

Mapping almost any work into a standard business functional organization converts it into the syntax of money language. It is very convenient—too convenient to totally abandon. Systems vary, but for example, processes for purchase orders, capital budgets, or placing ads are self-similar between organizations. People know how to communicate using transactions, and they can move from a business function job in one organization to a similar one in another, and within weeks become effective in the new setting.

The downfall of this language is that it translates almost everything into business models and marketing brands; houses, jobs, schools, water, jokes, funerals—even the sexual development of young girls.[35] That everything has a business aspect to it seems normal to those using this language. Seen from outside the system, it seems bizarre, and at times, repulsive.

Anything difficult to package, market, or financially value doesn't clearly map into this common business language. In recent years, the value of process attributes like flexibility, quality, and human development has opened wedges into this monetary syntax. But to those fixated on financial models, even something as simple as the strategic advantage of flexible operations is difficult to explain without falling into the black hole of money-speak from which they can't escape without learning a new vocabulary and syntax.[36]

To progress dealing with the challenges of Compression, the concepts of process flow, process interaction, mass–energy balance, and scientific problem solving, and even methods to deal with wicked problems, need to enter organizational language. To deal with Compression, we need to learn the language of natural and man-made physical processes. They don't speak finance.

Today's process language will modify with higher-tech communications, like Web 3.0, now in early development because languages are living, evolving systems. Intelligent software, searching millions of sites, could cough up things unavailable now. For instance, "bots" might report real-time data from remote sensors, so you can see wind speeds in a distant hurricane and instantly match that data with wind speeds from a prior hurricane at the same stage. If Web 3.0 develops as envisioned, human global language will become more compact and visual than written English.[37] But Web 3.0 is form. The substance is whether we can use it to help develop a common learning language to address critical questions like the following:

- Are we considering something truly important to our welfare?
- Will a change improve that welfare?
- Have we (or our systems) considered everything we can?

INABILITY TO SIMPLIFY COMPLEXITY

Many weaknesses of the current system of business arose because it promotes physical expansion and it can't stop; expansion is in the DNA of the system. By the latter twentieth century the system was also accumulating more process complexity than it could deal with. It is tightly interconnected, too. As can be seen in the financial meltdown, kicking out a prop in one part of the financial system has an unpredictable domino effect. And inside companies, financial officers are finding that the time and complexity of detailed budgetary controls exceed the value of using them by even their own cost measures; simplifying financial data gathering actually boosts financial results.[38] Regressing to an earlier stage to take a new direction (go bankrupt and start over) is not simple for the auto industry or most others. Few people can run back to live off the land again, an unspoken assumption of easy hire/fire policies a century or more ago, and centers of expertise, once disbanded, are not easy to reconstitute. People need transitional support so that simplifying systems to deal with Compression is not quick and easy. Our most basic assumptions and behaviors have to shift, reconstituting many of the work organizations we have today.

Commercial Competition; Commercial Reciprocity

Business culture has now penetrated many professions so deeply that it biases both their service to socially important missions and their professional learning. This commercial creep is often unseen by professionals infected by it. A few particulars may illustrate how "business think" advances quantity over quality when we need to reverse it.

In *Good Work: When Excellence and Ethics Meet*, authors Howard Gardner, Mihaly Csikszentmihalyi, and William Damon detail commercialism in two professional fields: genetics and journalism. Geneticists handle controversial science issues in stride, human cloning being one, but a survey reported in *Good Work* indicated that geneticists' most troublesome controversies were commercial conflicts of interest. Young ones were more troubled than older ones; doubly troubling because new entrants to a field generally learn its practical ethics from their mentors.[39] Unfortunately, the issues in genetics parallel those in almost all bioscience, including medicine.

Surveys of journalists indicated that they, too, fear business culture encroaching on their ethics. Investigative journalists probing controversial issues offend almost everyone at times. They expose many of the commercial conflicts in biomedicine, for instance. Journalists have a methodology, a code of ethics, and issues with journalists who do sloppy work, but feel undercut by business managers saving money by diluting news to grab viewers and boost advertising.

These fields contrast economically. Health care has grown rapidly. Traditional journalism feels threatened by extinction.

Health Care and Bioscience

Many leading academic biomedical researchers own companies or patents. Income from these side ventures sometimes dwarfs income from the university. State economic development programs eager to promote growth and jobs press researchers to start ventures, largely oblivious to the scientific issues: fear that researchers caught up in races to market sacrifice scientific rigor by excessive patent claims (as for genes), failure to disclose data, or shortcutting cross-validation. Colleagues that become competitors are less trusted to do the right things.

Industry funds only about 5 percent of academic research, but it attracts disproportional controversy.[40] Fine print in agreements may attach strings like rights of first disclosure or delays disclosing negative results. Industry influence takes on many forms. For instance, if all journal reviewers have industry ties, impartiality of the scientific review process becomes questionable. Review processes became so slow and dysfunctional that open source services now pry findings out of commercial lockups.[41]

The case of Dr. Hwang Wu Suk fabricating evidence of cloning human cells was so egregious that it made the popular news. Cases of commercialism corroding medical science usually don't, but they are so abundant that Web sites track them.[42] One of the better known ones is Alliance for Human Research Protection: http://www.ahrp.org.

In 2004, Marcia Angell and Jerome Kassirer (former editors of *The New England Journal of Medicine* now out of the heat) described how the slow drip of commercial reciprocity eroded professional obligations.[43] Angell concentrated on the pharmaceutical industry, Kassirer on physicians and researchers. Their books stood out in a crowd of similar writing because both were placed to see the state of medicine better than most people squirreled away in it, inclined to think things aren't really bad. I know no one involved; I haven't been bought; my professional judgment is uncompromised.

Since then, professional journals have tightened—a little. Authors and reviewers must disclose commercial ties, but incidents continue.[44] Public relations firms' medical communications specialists still ghostwrite for sign off by researchers.[45] Madison Avenue ad agencies invested in little companies that perform clinical trials and provide continuing medical education (CME), claiming not to influence research but merely speed new drugs to market.[46] However, despite adverse publicity, commercial CME is not free of influence.[47] Also well-known is that many clinical trials are poorly designed.[48] Industry has even funded academic research as fodder for PR claims.[49] Despite evidence that profit motives are not fully compatible with either social missions or research objectivity, industry executives protest that they are.[50]

Few people in the United States can miss pharmaceutical ads on TV. After the Food and Drug Administration relaxed stringent requirements to list all risks in the ads they proliferated, detailing (mostly pharmaceutical reps seeing doctors) increased from $3 billion in 1996 to $6.2 billion in 2002.[51] Seeing pharmaceutical reps in a doctor's waiting room is now common. Pharmaceutical companies did cut back on reps when some began going over the top.[52] Some reps even paid doctors to shadow them—observe in exam rooms without telling the patient who they were.[53] By adding drugs for enhancement besides drugs for treatment, drug consumption steadily rose.[54] Big Pharma gave stockholders the growth they loved, but critics accused it of "disease-mongering" to boost sales.[55]

Kassirer estimated that Big Pharma spends roughly $30,000 per physician per year promoting to them. Prescription drugs are only about 10 percent of total health care costs, and they prevent a great deal more expensive treatment, so Big Pharma growth prevents the rise in other costs—or so it is claimed.[56] But similar claims are made for most other procedures, tests, and devices sold to physicians by other companies. Physicians' prescriptions and orders trigger trillions in health spending, so sellers want to influence them, and payers (insurance companies) want to restrict them.

Big-margin blockbuster drugs are a big percentage of each Big Pharma company's revenue. Margins are preserved longer by extending blockbuster patents. The quickest way is by finding new "off-label" treatments (not yet FDA approved) for drugs now on the market, so subtle hints to doctors about off-label use have also stirred controversy.[57]

There's a lot more: jockeying with insurance companies to place drugs on their payment approval registries[58]; data-mining databases of doctors' lifetime prescription records[59]; lobbying the FDA and political leaders with the biggest lobby in Washington.[60]

Merck's Vioxx® is probably the best-known case of a company influencing clinical trial reporting, with considerable legal fallout.[61] The clinical trials of calcium channel blockers and ACE inhibitors to treat hypertension is a twenty-year-old story of many small studies with commercial claims, independent counterclaims, and charges of bias. It may not have concluded. One study found ACE inhibitors to be no more effective than diuretics (water pills), but 10 to 29 times more expensive. Finally a huge trial concluded that a calcium channel blocker given with an ACE inhibitor was more effective than a diuretic and its side effects were acceptable. During all this back and forth, one review of seventy journal articles found that 96 percent of supportive authors had financial ties to the drug makers; only 37 percent of negative-finding authors did.[62]

One wonders whether research on these drugs would have started were it not for the possibility of launching a blockbuster; but for any drug, sorting out evidence is hard enough without the pressure of commercial eagerness.[63] Busy pharmacists and physicians may not have time to pick through all the information available to them, so like the public, they too must depend on commercial competence as a public service.

Like Big Pharma, medical-device companies have come under fire for influencing physicians, although fewer cases have been reported.[64] Medical devices, like drugs, have sometimes gone awry. A clinical trial is never complete, and fail-safe processes and well-checked product designs are never reason to feel comfortable. An example is a Boston Scientific (formerly Guidant) problem with defibrillator leakage accompanied by the usual legal fallout.[65]

However, medicine is so complex that doctors have to depend on these companies for instruction and service on new devices. For example, *The New York Times* described a Stryker rep acting as a *bona fide* aide in an operating room, answering questions about Stryker knee implants, pulling a different size from inventory if needed—and monitoring Stryker instruments.[66]

Physician-corporate entanglements have taken strange twists. For example, an outspoken critic of Vioxx was Dr. Eric Topol, chief academic officer of the Cleveland Clinic Foundation. *Fortune* questioned his motives, noting that he was a paid consultant to a hedge fund betting on Merck stock to decline. To restore credibility, Dr. Topol ended his business relationships only to come under pressure from Cleveland Clinic management.[67] The Cleveland Clinic is a nonprofit and world-famous in research, but critics charge that its commercial ventures to fund its research both compromise its research and diminish its charity care to the uninsured poor.[68]

Expensive medical technology drives up health care costs, but so do quality issues with routine procedures (delay and rework). Third-party payment systems bloat administrative costs. Fragmented bureaucracies of specialists have difficulty communicating; the technology used for medical records is only part of the problem. The financial woes of big-city teaching hospitals with multiple missions, as in New York City, illustrate the scope and depth of the issues. Anyone with an eye for waste quickly spots opportunities. Almost everyone working in health care is dedicated to it, but the system doesn't serve its social mission very well. More isn't necessarily better.

Projections of current health care costs suggest that radical change must come sooner rather than later. The Congressional Budget Office projects that if nothing changed, by 2082 Medicare and Medicaid costs would be nearly 30 percent of total GNP, and total health care spending would be nearly 100 percent of GNP.[69] American medical technology is great, but overall, compared with other countries, the system is very expensive without making the population noticeably healthier. A favorite comparison of critics is that Cuba spends 7.3 percent of its GNP on health care; the United States spends 15.4 percent, but adult life expectancy is almost the same.[70]

Shannon Brownlee, medical industry analyst and author, has dug into many of the same health care issues much longer than I have and arrived at a similar conclusion. Monetary incentives misdirect medical processes and disrupt its critical human relationships.[71] (The malpractice bugaboo is merely the most politicized one.) For example, she cites overreliance on tests like CTs that do not aid diagnosis as much as physicians' think, but false positive CT diagnoses start an avalanche of further tests or treatments that snowballs both costs and risks.

More financial engineering, more marketing, more PR, and more lobbying clout can't fix this system; neither can going on ethics kicks, which do not dispel the basic conflicts of purpose built into the system or do much to simplify its complexity. Lean, quality, and patient safety initiatives help eliminate operational waste but do not dispel these systemic conflicts. Some of its incentives keep pushing health care expansion, quantity over quality, while cost-managing procedures removes responsibility from the people who deliver care without dulling commercial appetite to derive more revenue from everybody in the system—patients, providers, and payers.[72] Because the very ill will spend their life savings to find a cure, it's easy to find reasons why the health care system should keep expanding. And that system keeps shifting more to technical specialties than holistic physicians.

Change has begun, but it's slow. For example, Big Pharma's blockbuster pipeline is drying up while clinical trial costs soar north of a billion dollars a pop (depending on the accounting estimates). Lobbying the financial rules can't buy Big Pharma much more time.[73]

Of the alternative pathways, one that seems likely is bioservices that predict treatment efficacy based on genetic codes, designing a molecule or other intervention to go exactly where needed for a specific individual, perhaps with a nanobot carrier.[74] This technology is in infancy and has many hitches, but it already threatens Big Pharma's old industrial model: producing medicaments in big batches to sell by the dose. Eventually big statistical clinical trials will be displaced by personal monitoring and intervention—clinical trials in a lot size of one, a patient, who would receive a customized intervention. That's many years away, but it suggests the future of health care will be much more preventive and anticipative. Of course, doing this takes investment capital, but the key is highly skilled people using a highly developed continuous learning system.

Politically and ethically, health care involves contentious, social decisions: When should heroics to save a life stop? When should we regard a blob of protoplasm as a new human life? What should we do with people who abuse their own health and that of others?

Public policy may decide who has access to health care and how, but it can't fix a health system at the green-gown level.[75] Stronger medicine is needed—a bottom-up revolution to continually improve work processes and work organization to better deliver what patients need. Focus on mission: improving quality of health and quality of health care practice, not creating something to sell. To do this, everyone has to engage in holistic process learning; technical advancement is not enough.

Commercial Journalism

Health care has been a growth industry, but journalism, or at least traditional news media, have been shriveling and cost cutting for years. Journalists are of interest because they have had difficulty defending their concepts of reporting as a public service from interlopers and from encroachment by managers running the business.

Back in 2000, *Good Work* found a higher percentage of journalists than geneticists troubled by the decline of their own professionalism. They were upset by (1) pressure to advance their employers' business goals, and (2) lack of journalistic values and ethics. Field journalists were so much more troubled than media managers that the surveys appeared to describe two different populations.

School-trained journalists have lapses, of course. Many lapses are reported by *Editor & Publisher*, and if egregious, by the general media. If well done by the code of ethics, journalism practice follows a fairly rigorous structure, and doing so may be both intellectually and emotionally taxing. A few excerpts from that code are in the bullet list on the following page. The full code, detailed and pages long, can be seen at http://www.spj.org/ethicscode.asp.[76] However, what journalists fear is faux journalists with no concept of the code, including bloggers. Since 2000, bloggers bringing down prominent journalists have increased fear that political bias on top of commercial creep threatens objective journalism.[77] Many bloggers are self-proclaimed journalists with no concept of its standards or ethics. Through outreach programs, schools of journalism are attempting to take corrective action.[78]

Journalists should:

- Test the accuracy of information from all sources and exercise care to avoid inadvertent error. Deliberate distortion is never permissible.
- Diligently seek out subjects of news stories to give them the opportunity to respond to allegations of wrongdoing.
- Identify sources whenever feasible. The public is entitled to as much information as possible on sources' reliability.
- Be free of obligation to any interest other than the public's right to know.
- Avoid conflicts of interest, real or perceived.
- Be wary of sources offering information for favors or money; avoid bidding for news.

Looking at the preceding lists, reflect on how you would personally live up to this, and then on the parallels of the code with scientific problem solving (like Plan-Do-Check-Act). Understanding a complex story clearly enough to condense its essence in limited time or space is an intellectual challenge. And presenting all known, relevant facts of a controversial story in an unbiased way while being pressured to do otherwise is a test of courage.

My personal experience is that many business folk don't know journalistic ethics. If intent on promoting their company (or a cause), they assume that media are opportunities to look good. Commercial reciprocity being how things are done, they assume that if they make nice and pay expenses, coverage will be favorable. If not, they are disappointed. If reported facts are wrong or incomplete, they are really upset. Then they either avoid media contact, or hire PR representatives who understand news systems and cycles. Of course the PR representatives' job is to favorably influence the news.

Commercial influence managing news operations is not outright censorship, but more subtle. It's amoral in the pursuit of a bigger audience.

In the opinion of American journalists, corporate ownership and commercial influence is the root of the news rot. Owners install media business managers to make money. They don't overtly intervene in reporting. They influence it through marketing and budgets. Identify market segments advertisers want to reach. Budget whatever attracts those segments—sports, entertainment, local dining. Lop off anything that doesn't. If that blurs news into entertainment, let fewer reporters ask nosy questions, and yuk up banality to hold eyeballs between ads; well, you have to present whatever makes money.[79]

Some journalists passionately resist commercialization; others go with the flow. Even staid old BBC sustains revenue for news by hawking cuddly toys to enhance its brand image and designing British programs to become American hits. Despite such revenue pumping, in 2005 the Beeb cut its total staff by 2050 people.[80]

Blatant clashes between journalism and commercialism are rare. Commercial influence is subtle, persistent, and almost imperceptible to business thinkers. For example, in the October 10, 1999, issue of the *Los Angeles Times*, a 168-page Sunday magazine special promoting the new Staples Center described the center and its tenants and was loaded with advertising. Unbeknownst to those writing this issue, *Times* publisher Kathryn Downing had agreed to divide its advertising revenue with the Staples Center. As a business strategy, this seemed perfectly logical; as a journalistic practice, it compromised the integrity of reporting.

On Monday, December 20, 1999, the *L.A. Times* published a 14-page investigation in which media critic David Shaw laid out what happened and why, concluding that the newspaper's integrity was compromised if *any* story it published had hidden connections and payoffs. If it had happened once, how could readers trust any other story written about the Staples Center or about any other organization?[81]

Shaw reported that Kathryn Downing had apologized, pleading ignorance of journalistic practices (in my experience, few other executives would either). Michael Parks, primary editor for the special issue, couldn't remember what happened. Statements of principles were issued. That did not pacify Shaw. He questioned whether commitment to ethical journalism would last longer than the next bottom-line decision. Few newspapers publish their ethical guidelines, if they have any.[82]

In 2000, Times-Mirror sold the *L.A. Times* to Tribune, the media conglomerate that publishes the *Chicago Tribune*. *Los Angeles Times* journalists then questioned the long-term commitment to journalism of the Tribune company, which also owns the Chicago Cubs, 23 TV stations, and other media. Relations between Tribune management and the *L.A. Times* news operation became tumultuous—changing editors, protests, cutbacks. These clashes were the most visible of many traditional news operations battling to uphold standards while newspaper managements looked for imaginative ways to hold readers and advertising.[83]

Between 1980 and 2002, the number of newspapers in the United States declined 17 percent, and it keeps declining. Younger people who are more wired pay scant attention to hard news from any media.[84] Subscription revenue hardly covers its collection costs. Because ads pay the bills, space devoted to ads grew and grew. To battle this, newspapers converted from broadsheet to tabloid style, posted Web pages, and went for quicker reads, grittier news, and more come-ons. But is this competition for quantity killing quality journalism in order to save it?

When TV was new, it was naively hoped that it would increase the scope and depth of public knowledge. However, hard news and education lack the attraction of entertainment.[85] Newscasts mimicking entertainment hoped to rivet audiences in coveted market segments. With few exceptions, it didn't work.

Concentration of media ownership is an issue. Business thinkers regard it as an issue of market monopoly, locally or nationally. Journalists regard it as blandness and cost-control undermining serious, independent reporting. To draw audience, mass media chains run news like fast food—too quick to serve anything complex, and too bland to upset either sponsors or users. Independent news sources, from Investigative Reporting Workshop to ProPublics.org, arose, but most also have financial woes. With media thus splintering, people are apt to gravitate only to sources that fit their biases.[86] That deprives society of unifying, objective sources of news on anything controversial.

News as presented by Clear Channel is an example critics use of a Big Media monopoly watering down or even distorting news. Clear Channel owns about 1,200 U.S. radio stations and 240 foreign ones, an entertainment business so aggressive in cost cutting that critics nicknamed it "Cheap Channel."[87] Its creed reflects business ethics more than journalistic ethics.[88] Political critics accused Clear Channel of even staging events to liven up the news, for example organizing pro-war rallies during the run-up to the Iraq war.[89]

Not only print newspapers but all "big old media" have had to chop the numbers of field reporters, amid rising concern for exacerbating already-weak quality of coverage.[90] For example, TV stations' news rooms short of staff turned to outside help like video news releases (VNRs). These are video clips on whatever, usually made by PR companies for any client wanting media exposure, and often spliced into regular news broadcasts without disclosing the source. Almost everyone can identify an undisguised VNR, a late-night TV infomercial touting a $19.95 special.

However, the less difference that can be detected between commercial persuasion and neutral reporting, the more persuasive, so imaginative PR is indistinguishable from objective media reporting. Even leading journalists have succumbed to blandishments to trade on their credibility plugging products and services. For example, pharmaceutical and health care companies enticed Aaron Brown of CNN and Walter Cronkite, the retired CBS news anchor, to appear in promotional videos (Cronkite backed out).[91] *The News about the News* sums up the conflict thusly: "Covering the news, once seen as a public service that could also make a profit, became primarily a vehicle for attracting audiences and selling advertising, to make money."[92]

Spin Doctoring

The problems of professional journalism, at least as an ideal, are not understood without sensing its conflict with PR seeking to influence public perception to favor clients, much like lawyers with the legal process. PR seeks to implant in the public mind ideas or images that may be factual—or may not. A real win reshapes the public mind. PR's original spin doctor, Edward Bernays, knew it was a science of propaganda, playing to emotion, not logic, to "engineer" the human psyche.[93] Neuroscientists mapping the brain's decision-making process sparked "neuromarketing": triggering a buy decision just by stimulating the right neural reflexes, which is more marketing than science so far.[94]

PR campaigns are a long-term sequence of change, based on the "sciences" of semantic transfer, image creation, and often, outright deception—like a political campaign. The Reagan administration was once described as a "PR outfit that became president and took over the country."[95] Conventional political advice became that without PR money, you don't have a chance in major campaigns, so they became PR wars. Political leadership has always been theatrical, but today's political PR advisors are apt to think of a campaign as a business. The goal is to win, and as in professional sports, the franchise with the most money has an edge.

PR professionals have a code of ethics that critics call an oxymoron, but some practitioners do try.[96] For example, to PR consultants like Peter O'Malley, serving any paying client isn't ethical: "If we are ethical as PR practitioners, it means we choose to serve clients whose self-defined interests are, in our view, ethical. Or we clear out. Period."[97] Crisis management counseling, like Johnson & Johnson's episode of Tylenol terrorism, is PR's perennial example of their work at its best. Examples of helping clients cover up or mislead aren't touted.

Competing PR firms work different sides of public issues, each striving to win, mining facts and data for persuasive nuggets. Three favorite PR weapons of choice are:

- VNRs that a TV station or Webcaster may use without disclosing its source to the public. Commercial companies, nonprofits, and government agencies have all used VNRs since the 1980s.[98]
- Training organization officials to respond to media inquisitions during crises: Either lay it all out, or say as little as possible in as many words as possible.
- Astroturf grassroots campaigns: Set up misnamed groups. Infiltrate established ones. For example, the Texas Beneficial Use Coalition was concocted to inveigle Texas and the townsfolk of Sierra Blanco into landfilling sewage sludge from New York City.[99] Likewise, set up authoritative-sounding think-tanks to draw media attention away from real researchers on various issues.

PR battles spew disjoint concepts, often to promote a sales campaign. For example, a puffed report of a new dietary study, if not anchored in context to other findings, may launch a new diet fad.

The American public is inured to commercial propaganda—ads—but may be surprised at how many government agencies use VNRs to promote various programs like vaccination campaigns. Few of these stir angst among journalists unless they muddle political reporting, like Pentagon "news operations." Examples are paying Iraqi journalists to write favorable stories in 2005; following up with news Web sites in foreign locations from Iraq to South America[100]; and managing retired generals as expert commentators on war-related issues.[101] Journalists charge that the Pentagon can't report on itself and that it is undermining not only its own credibility, but that of independent reporting in areas where government-controlled media already have little credibility.

Both in Washington and on Madison Avenue, PR is just another job, as with Karen Ryan, who narrated "news releases" as if she were a reporter.[102]

Only those who claim to be professional pay a penalty. For example, the commentator Armstrong Williams was paid $240,000 to narrate VNRs promoting the Department of Education's No Child Left Behind law. When *Tribune* editors discovered this, they fired him.[103] "News management" for political aims stirs more controversy than that for commercial aims.[104]

PR is amazingly persuasive, even on people aware of logic traps. When we know few facts of a matter, we are apt to trust a self-assured blowhard over a professional cautiously weighing mixed evidence. That is why physicians take on that confident air easily copied by actors. PR is almost the opposite of scientific reasoning. It fools us into confounding reality with images, like a Yanomamö shaman mixing real observations with his conjured images. A PR "win just to be a winner" is a modern version of medieval trial by combat. If done only for a payoff, there is no questioning of whether society is the better, which is the opposite of professional dedication and scientific objectivity.

PR at its worst is a media version of self-promotion, and most business executives need to promote something. They don't intend to deceive. They merely bullshit to advance their objective—scoring a win or a sale. Veracity or lack of it is irrelevant to their goals. "Bullshit artists" honestly present themselves and their aims with great self-assurance, cutting off other people in meetings, countering evidence by spinning conclusions in reverse, and exhibiting the behaviors shown in the list below.[105] (Carl Sagan covers many more of these in *The Demon-Haunted World: Science as a Candle in the Dark*, Random House, New York, 1996. We can resist these demons, but no human can escape them.) For Americans, the fears of propaganda experts are politically ominous: "In the long run, our seemingly insatiable desire for entertainment may succeed where Hitler and Pravda failed."[106] But from a commercial view, is believing your own PR the final stage of self-deception when a system is dying of its own incompetence?

- Ad hominem: Attack people personally without checking their observations or logic. "Joe is always complaining, trying to get out of work."
- Deference to authority: "The president must have a reason we don't know about." "The market is never wrong." "That's what Dr. Deming said himself."
- Claim that an issue is beyond the subject's understanding: "You don't have the big picture of top management." "It's a complicated tax issue."

- Selective observation: Cite data that supports your position. Omit, or even suppress, any that does not.
- Argument by anecdote: Tell a story that supports your position. Generalize it to cover all cases. Ignore any anecdotes to the contrary.
- Stonewall: Claim that data or processes are secret or proprietary, do not exist, or that digging it up is too costly—adult versions of "the dog ate my homework."
- Don't question an assumption (related to denial and begging the question): "The Titanic is the most advanced ocean liner ever built; unsinkable."
- Coincident timing or correlation is assumed to be causation: "Sales of whiskey in Oxford correlate 0.85 with reported faculty salaries, so...."
- Conspiracy theories: A "group out to get us" expands ad hominem beyond one person. "The (fill in the group) must be paying (whistle-blower) to cause trouble."

Commercial Conflict and Reciprocity

Expansionary business is more like a gold rush than life on a space sovereignty, or in growth-constrained preindustrial societies. In a rush for riches, waste is of little concern, and winners can consume sumptuously until it is all gone. However, even to gold miners, quality performance from both health and news sources is important at present. In Compression that quality of performance needs great improvement, but as illustrated by both cases, business incentives confound dedicated service to society by professionals who may have taken an oath to do so. Business logic tends to stress quantity over quality both with pressures to expand revenue (health care) and by pressures to cut costs (journalism). To build quality, leadership has to continuously resist that logic.

Business logic infects professionalism through little cuts of "commercial reciprocity": a little money for helping others make money. That logic justifies investors giving bonuses to executives for running up a market cap, giving physicians fees to promote medical devices, and giving journalists fees for promoting a cause. It's a variation on the logic that everyone working in self-interest will benefit the whole. But if people have to share resources and rapidly learn together to survive, this logic begins to break down. It undermines the necessary level of trust.

QUESTIONING THE SYSTEM

Compression changes everything. Global survival dwarfs the competitive interest of any company or country. Individual human capabilities must improve. Work organization effectiveness must dramatically improve. The dominant basis for status must shift from *getting* to *doing*; from "management for results" to "leadership for learning."

Nothing in the experience of business leaders—or almost anybody—has prepared us for this, so we stay in denial. Surely a few shortages or environmental issues are mere hiccups, after which industrial society will continue expanding. Within anyone's living memory, it always has.

To lift thinking out of business-as-usual ruts, consider some mind-bending questions. Assume that you are in a space sovereignty like the scenario that began the chapter:

Would it use money? If so, what for?
Why measure the time value of money? How? For what?
What would "investment" mean?
What would work consist of? How would it be organized?
How would people be rewarded for work?
What would be defined as an innovation?
How would people generate new ideas? How could they try them out?
    How could they foster creativity while avoiding "group think"?
How would they know whether the performance of a life-critical system
    had been improved? How would they identify problems and decide
    what to do about them?
What would be the governance system?
How would they evaluate individual performance?
How would they develop people to assume future positions of
    responsibility?

Of course, earth's population of nearly seven billion being much bigger than the space sovereignty, leading humanity through such mind-bending change is mind-boggling—impossible by dictatorial fiat. People change when they want to, or see that they must. Will we revert to Stone Age chaos or evolve a higher form of civilization? If it can be done, how can we start? By creating very different kinds of work organizations.

Creating new work organizations, or changing existing ones is daunting, but at least it has been done. Large work organizations are so unwieldy that leaders may need to break them into moldable pieces, and even small ones can't evolve a new paradigm in a few months. And few will be able to do all this without being supported by a learning network, or supply chain including financiers. So how can we do this? For starters, ask yourself a few more questions, like:

> How can companies embrace social missions if subject to the short-term demands of global capital markets? What has to change?
>
> Can most people become professionally dedicated to a social mission?
>
> Without using money how can an organization know whether its performance is improving?
>
> How can all people in an organization individually take on more responsibility while coordinating their collective effort?
>
> How can an organization establish a social mission that inspires people more than monetary incentives?
>
> How can an organization create a learning system with a high density of communication?
>
> What kind of leadership behavior devolves maximum, genuine responsibility on those doing the work?

Toyota did not become a paragon of proficiency by *behaving* as a normal profit-maximizing company. Despite all the advantages that the Toyota Way conferred, it competed for market expansion and is now caught in the same trap as the rest of the vehicle industry. Compared with what is required to cope with Compression, Toyota's changes have barely started. Addicted to its current business model, Toyota has barely begun to think about changing it to retrofit low fuel-economy vehicles now in service for example.[107]

The legacy business system cannot deal with Compression. How to meet its challenges is hard to comprehend from an expansionary mind-set, for dealing with Compression is not a competitive game, but a collective challenge. Work must consist less of selling more stuff and more of learning what to do. Nobody knows how to meet these challenges. We have to learn, and some lessons are tough ones.

Dealing with Compression—the realities of nature, resources, and ourselves—is no minor tweak to capitalist economics. Sugar-coating that reality in familiar business language so that the business community will get it soon reaches a limit. It has to learn a different language. That is why we need vigorous learning enterprises—to learn it quickly.

ENDNOTES

Extended version of endnotes available at http://www.productivitypress.com/compression/footnotes.pdf.

1. Gerard K. O'Neill's "Island Three" design in *The High Frontier*, William Morrow, New York, 1977. The space colony scenario is a smaller variation of his third-generation "Island Three" design.
2. This problem is illustrated by complex protein folding models; digital computing time can exceed the age of the universe.
3. A thick wall plus air inside attenuates radiation sufficiently for astronauts to survive a few months near earth. Long-term radiation effects on multiple generations in deep space are unknown.
4. Near-earth spacecraft now use 101.3 kPa (about 15 psi) atmospheric cabin pressure with 21 percent oxygen, and the shuttle uses 70.3 kPa, with elevated levels of oxygen. See NASA/CR-2005-213689.
5. Patrick L. Barry and Dr. Tony Phillips, demonstrate the Coriolis effect on a NASA site: viewable at http://orc.ivv.nasa.gov/vision/space/livinginspace/23alvl_spin_prt.htm
6. Advanced versions of O'Neill's Island Three assume a nearby sun, slower rotation, and replication of a daily solar cycle using mirrors and closable windows to offset the vertigo of humans spinning in starlit emptiness.
7. See Mark Prado, PERMANENT; http://www.permanent.com/s-centri.htm.
8. Eugene N. Parker, "Shielding Space Travelers," *Scientific American*, March 2006.
9. William J. Broad, "Orbiting Junk, Once a Nuisance, Now a Threat," *New York Times*, Feb. 6, 2007. The amount of debris orbiting earth has increased since this article.
10. The history of Biosphere 2 plus ongoing, related research is at http://www.biospherics.org.
11. Albert A. Harrison, *Spacefaring: The Human Dimension*, University of California Press, Berkeley, 2002.
12. During the twentieth century, dramatic human modification with implants, prosthetics, and drugs became rather prosaic. See the World Transhumanist Association (http://www.transhumanism.org) and Chapter 6.
13. The Space Foundation's Web site is http://www.space-frontier.org. Possibilities having commercial promise are at http://www.cygo.com/space_products.html.
14. For example, Todd Lewan, "Chipping at Privacy?" *Chicago Sun-Times*, July 22, 2007.
15. Robert Kaplan and David Norton, "The Balanced Scorecard: Measures that Drive Performance," *Harvard Business Review*, Jan.–Feb. 1992.
16. PACE Awards are given annually to automotive suppliers for outstanding innovation. (The author is a judge.) Responsible Shopper is a program of Green America (www.coopamerica.org) Great Place to Work Institute is at www.greatplacetowork.com
17. Brian H. Maskell and Bruce L. Baggeley, "Lean Accounting: What's It All About?" *Target* (publication of the Association for Manufacturing Excellence), Issue 1, 2006.

18. "Accuracy" is not the best term to describe any accounting system, because none really have a standard of comparison to measure against, like a standard reference for measuring distance.
19. Roger Lowenstein, *When Genius Failed: The Rise and Fall of Long-Term Capital Management*, Random House, New York, 2000.
20. Figures were taken from the Commodity Futures Trading Corporation per *Bloomberg News*.
21. "Carbon Markets Create a Muddle," *Financial Times*, April 26, 2007. A summary of these markets' "success" is Mark Scott, "Europe's Carbon-Trading Pioneers," *Business Week*, April 28, 2008.
22. Herman Daly, *Beyond Growth*, Beacon Press, Boston (paperback), 1997.
23. Juan Williams, "The Faithful's Wayward Path," *New York Times* op-ed page, January 20, 2003.
24. Kalle Lasn, *Culture Jam: How to Reverse America's Suicidal Consumer Binge—and Why We Must*," Perennial, New York, 2001.
25. Gerald Wilde, *Target Risk*, PDE Publications, Toronto, 1994 (the concept of risk homeostasis).
26. Carpet design for recycling is not trivial, and Interface addressed issues that would never be addressed if waiting for the "market" to demand that they be addressed.
27. Marjorie Kelly, *The Divine Right of Capital*, Berrett-Koehler Publishers, Inc., San Francisco, 2001.
28. Noreena Hertz, *The Silent Takeover*, The Free Press, New York, 2001.
29. David C. Korten, *The Post Corporate World*, Berrett-Koehler, San Francisco, 1998, p. 187.
30. Mark Bowden, *Killing Pablo*, Penguin Books, New York, 2001, pp. 28–29.
31. Jeremy Hope and Robin Fraser, *Beyond Budgeting*, Harvard Business School Press, Boston, 2003. In the United States, Beyond Budgeting centers on the Beyond Budgeting Round Table: http://www.bbrt.org.
32. See for example, Philip Kotler and Nancy Lee, *Corporate Social Responsibility*, John Wiley & Sons, New York, 2004. A typical rebuttal is Betsy Atkinson, "Is Corporate Social Responsibility Responsible?" *Forbes*, Nov. 28, 2006.
33. Paul Maidment, "Re-Thinking Social Responsibility," *Forbes*, Jan. 28, 2008.
34. The Web site for Corporation 20/20 is http://www.corporation2020.org
35. M. Gigi Durham, *The Lolita Effect*, The Overlook Press, New York, 2008 (on the consequences of marketing to girls aged 8 to 12 to look as hot as Paris Hilton or Britney Spears).
36. Stephen A. Ruffa, *Going Lean*, AMACOM, New York, 2008.
37. Picture symbols are older than phonetic alphabets, but how do you text message with them? Electronic communication is changing language.
38. This refers to the Beyond Budgeting movement, cited earlier: http://www.bbrt.org.
39. Howard Gardner, Mihaly Csikszentmihalyi, and William Damon, *Good Work*, Basic Books, New York, 2001.
40. The percentage is from *Science*, May 14, 2008, p. 1549.
41. David Cohn, "Open Source Biology Evolves," *Wired*, Jan. 17, 2005. A later *Wired* article (April 11, 2005) reported that use was growing rapidly.
42. Julie Bosnan, "Reporters Find Science Journals Harder to Trust; Not Easy to Verify," *New York Times*, Feb. 13, 2006.
43. Marcia Angell, *The Truth About the Drug Companies*, Random House, New York, 2004; and Jerome Kassirer, *On the Take*, Oxford University Press, Oxford, U.K., 2004.

44. Lawrence K. Altman, "For Science's Gatekeepers, a Credibility Gap," *New York Times*, May 2, 2006.
45. Shannon Brownlee, "Doctors without Borders," *Washington Monthly*, April 2004.
46. "Madison Ave. Plays a Growing Role in Market Research," by Melody Petersen, *New York Times*, Nov. 22, 2002.
47. As of 2008, almost all the 741 registered CME providers by the Accreditation Council for Medical Education were recognizable medical schools or institutes, but one could still find market research on "leveraging CME as a channel of influence."
48. Dr. James N. Weinstein, editor-in-chief of *Spine*, in 2006 proposed a National Clinical Trials Consortium to overcome the obvious problems with trials all paid by companies highly desirous of a positive outcome.
49. Stephan Heres et al., "Why Olanzapine Beats Risperidone, Risperidone Beats Quetiapine, and Quetiapine Beats Olanzapine," *American Journal of Psychiatry* 2006 163:185–194. This widely cited article noted that in most cases, pharmaceutical funded studies simply omitted from trial any drug that might show clearly superior performance.
50. Business logic eroded Big Pharma's mid-twentieth-century public service obligation that every ethical house should carry a full line in the interest of public health, whether profitable or not, as illustrated by the complicated story of vaccines. Financial reporting rules plus risk of obsolescence in storage nearly dried up national vaccine stocks.
51. The Henry J. Kaiser Family Foundation, *Trends and Indicators in the Changing Healthcare Marketplace, 2004 Update*, April 2004: Pharmaceutical promotion data are from Exhibit 1.20.
52. Jamie Reidy, *Hard Sell*, Andrews McMeel Publishing, Kansas City, MO, 2005.
53. Kassirer, op cit., p. 49.
54. David Amsden, "Life: The Disorder," *Salon*, Nov. 25, 2005.
55. Ray Moynihan, Iona Heath, and David Henry, "Selling Sickness: The Pharmaceutical Industry and Disease-Mongering," *British Journal of Medicine*, Apr. 13, 2002, 324:886–891.
56. U.S. prescription sales grew to $286 billion in 2007, about 10 percent of total health care spending. Generics were over half of all prescriptions, but only about 20 percent of the market revenue.
57. Alex Berenson, "Lilly E-Mail Discussed Off-Label Drug Use," *New York Times*, Mar. 14, 2008. The company has posted its position at http://newsroom.lilly.com/ReleaseDetail.cfm?ReleaseID=299792.
58. Popular prescription guides, such as *The Pill Book*, may be dated. Professional physicians and pharmacists use up-to-date registries, but drug companies attempt to influence them. Kassirer, op. cit., cites a case of this.
59. Robert Steinbrook, M.D., "For Sale: Physicians' Prescribing Data," *The New England Journal of Medicine*, June 29, 2006.
60. According to both the Center for Public Integrity and the Center for Responsive Politics.
61. Alex Berenson, "Revamping at Merck to Cut Costs," *New York Times*, Nov. 29, 2005.
62. Melody Petersen, "Diuretics Value Drowned Out by Trumpeting of Newer Drugs," *New York Times*, Dec. 18, 2002.

63. "A Cure for the Common Trial," *Science*, May 12, 2006. The Public Library of Science's *PLoS Clinical Trials*, is publishing all sound clinical trials regardless of outcome, but can those who need to know wade through all of it?

64. Reed Abelson, "Hospitals See Possible Conflict on Medical Devices for Doctors," *New York Times*, Sept. 22, 2005.

65. Barry Meier, "Repeated Defects in Heart Devices Exposes a History of Problems," *New York Times*, Oct. 20, 2005; and Doug Bartholomew, "Quality Takes a Beating," *Industry Week*, March 2006.

66. Barnaby J. Feder, "A Parts Supplier to an Aging Population," *New York Times*, March 26, 2005.

67. Reed Abelson and Stephanie Saul, "Ties to Industry Cloud a Clinic's Mission," *New York Times*, Dec. 17, 2005; and Andrew Pollack, "Medical Researcher Moves to Sever Ties with Industry," *New York Times*, Jan. 25, 2005.

68. Reed Abelson and Andrew Pollack, "Patient Care vs. Corporate Connections," *New York Times*, Jan. 25, 2005.

69. Senate testimony of Peter R. Orzag, director of the Congressional Budget Office, Jan. 31, 2008, Table 4, found at http://www.cbo.gov/ftpdocs/89xx/doc8948/01-31-Health Testimony.pdf.

70. Comparisons derived from Annex Tables 1 and 2, World Health Organization, *World Health Report 2006*. Cuban medicine is louded for basic care for everyone, but criticized for "medical apartheid" in the best facilities.

71. Shannon Brownlee, *Overtreated*, Bloomsbury USA, New York, 2007.

72. Elliot S. Fisher and H. Gilbert Welch, "Avoiding the Unintended Consequences of Growth in Medical Care," *Journal of the American Medical Association*, 281, 1999, pp. 446–453.

73. Alex Berenson, "Tax Break Nets Drug Firms Billions, *International Herald Tribune*, May 9, 2005.

74. A general overview of research on the "diseasome" is Andrew Pollack, "Redefining Disease, Genes and All," *New York Times*, May 6, 2008.

75. Jan Hoffman, "Awash in Complexity, Patients Face a Lonely, Uncertain Road," *New York Times*, Aug. 14, 2005.

76. Daniel Shorr, column in *The Christian Science Monitor*, May 16, 2003.

77. Kathryn Q. Seelye, "Bloggers as Media Trophy Hunters," *New York Times*, Feb. 14, 2005.

78. Rory O'Connor, "Journalism Drowning," MediaChannel.org, June 1, 2005.

79. Ted Koppel, "And Now, a Word for Our Demographic," *New York Times* guest editorial, Jan. 29, 2006.

80. Heather Timmons, "Britain: More Cuts at BBC," *New York Times*, Mar. 22, 2005.

81. David Shaw, "Special Report: Crossing the Line," *Los Angeles Times*, Dec. 20, 1999, p. 1.

82. *Al Jazeera* issued an extensive ethical statement after Western criticism in 2004. It can be found in the "about us" section of the home page.

83. Tribune financial statements (at http://www.tribune.com) take research to decipher, but in them one can see the confusing clashes between public service commitment, being "competitive," and serving the capital markets.

84. Joan Shorenstein Center at Harvard, "Young People and News," which can be found in the research reports at http://www.hks.harvard.edu/presspol.

85. Vance Packard, *The Hidden Persuaders*, Random House, New York, 1957.

86. Ironically, Independent Press Association, formed in 2000 to assist small publication distribution, failed in 2006, largely because it could not manage chaotic finances. A similar group is still going in New York City.

87. Eric Boelert, "Radio's Titan Hits the Skids," *Salon*, Aug. 7, 2002; this is one of 34 stories in *Salon* on Clear Channel, helping fuel congressional pressure on the FCC not to expand the scope of media ownership.

88. Can be found at http://www.clearchannel.com under "About Us".

89. Christine Y. Chen, "Not the Bad Boys of Radio," *Fortune*, Mar. 3, 2003.

90. Richard Pérez-Peña, "Big News in Washington, but Far Fewer Cover It, *New York Times*, Dec. 17, 2008.

91. Melody Petersen, "A Respected Face, but Is It News or an Ad?" *New York Times*, May 7, 2003.

92. Leonard Downie, Jr., and Robert G. Kaiser, *The News about the News*, Vintage Books, New York, 2002, p. 243.

93. Larry Tye, *The Father of Spin: Edward L. Bernays and the Birth of Public Relations*, Owl Books (Henry Holt), New York, 1998.

94. Randy Dotinga, "Advertisers Tap Brain Science," *Wired News*, May 31, 2005 Neuroscientists regard this idea as hype way beyond the present state of knowledge.

95. Mark Hertsgaard, *On Bended Knee*, Schocken Books, New York, 1989, p. 6.

96. Public Relations Society of America: http://www.prsa.org/aboutus/ethics.

97. Peter O'Malley, "In Praise of Secrecy," prior column on All about Public Relations Web site. The message which was quoted is now taken down, but O'Malley's messages are very revealing.

98. In 2006, The Center for Media and Democracy tracked thirty-six VNRs out of thousands that had been distributed; seventy-seven TV stations reaching about half the U.S. population had used one of these thirty-six in a newscast without revealing that it was not their own reporting: http://www.prwatch.org/fakenews/execsummary.

99. John Stauber and Sheldon Rampton, *Toxic Sludge Is Good for You*, Common Courage Press, Monroe, ME, 1995, p. 117.

100. Peter Eisler, *USA Today*, May 1, 2008. The Pentagon's site is at http://www.mawtani.com/ar.

101. David Barstow, "Behind TV Analysts, Pentagon's Hidden Hand," *New York Times*, Apr. 20, 2008.

102. David Barstow and Robin Stein, "Under Bush; A New Age of Pre-Packaged Television News," *New York Times*, Mar. 13, 2005.

103. Dave Astor, "Armstrong Williams Column Axed by TMS," *Editor & Publisher*, Jan. 7, 2004.

104. Dave Astor, "*Business Week's* Javers Hints that More Paid Pundit Stories May Be Coming," *Editor & Publisher*, Jan. 18, 2006.

105. Harry G. Frankfurt, *On Bullshit*, Princeton University Press, Princeton, NJ, 2005.

106. Anthony Pratkanis and Elliott Aronson, *Age of Propaganda*, revised ed., Henry Holt, New York, 2001.

107. Toyota's Annual Environmental Report is at http://www.toyota.com/about/enviro report 2008.

5

Creating Vigorous Learning Enterprises

Forming a vigorous learning enterprise is counterintuitive to much normal behavior in expansion. Conventional businesses are financially controlled, with status systems, hierarchies, and politics based on money and control. Lean or Six Sigma initiatives barely dent this behavior, especially if business minds regard them as cost-cutting techniques, but leaders of these initiatives frequently describe them as eye-opening life experiences. Work organizations capable of taking on the challenges of Compression are a major leap beyond.

Just holding all the challenges in Chapter 1 in mind at once is a huge challenge in itself. By addressing only one at a time, we consider making substitutions within the current system, for example, displacing all fossil fuels with alternatives.[1] But if it comes to that, people will not long sacrifice food for fuel. Considering all challenges at once, we enter Compression: Assure survival of life and promote quality of life using processes that work to perfection with self-correcting, self-learning systems. No use of excess resources. No wasted energy. No toxic releases. Quality over quantity, always.

Arbitrary quantification sharpens this: Worldwide, create at least the same quality of life as in industrial societies today, while using less than half the energy and virgin raw materials, and cutting toxic releases to nearly zero. To do this, high-consumption areas have to cut much more than areas where people can hardly survive now. Nobody knows how much Compression is sufficient, but it's not trivial. Grand-scale objectives won't be met quickly, so how can we begin on a small scale to start moving in this direction?

Creating vigorous learning enterprises is a start, but it takes more than organizational restructuring, new technology, and new programs. We have to learn how to think differently, individually and collectively. Instead of racing to grow, race to shrink. Instead of envying conspicuous consumers, regard them as wastrels and profligates. See problems in a different way and concentrate imagination on them. If we have the will, we have capability to accelerate our technical progress.

This will not be easy because industrial society has worked itself into a "progress trap"—a state from which it is difficult to escape quickly.[2] Any way we wiggle bumps into what is known as "wicked problems" (imaginary complexity because we can't give up the past to agree on facts or what to do). None of us can learn how to emphasize quality over quantity in a few days.

To start this change, create a different kind of work organization, superior to anything today: differently led, differently motivated, and with different patterns of thought—a vigorous learning enterprise. Experiment with small work groups, and improve the ideas for these organizations. Their missions may correspond to nonprofit, for-profit, governmental, or hybrid organizations today. The infrastructure they create and the thought leadership they exhibit will do much to coach the general population to cope with Compression. Expand them to become global in scope. Some may teach have-nots how to improve their quality of life in a resource-short world.

VIGOROUS LEARNING ENTERPRISE OVERVIEW

No prior business language conveys the concept of a vigorous learning enterprise. If the ideas enter practice, eventually a short name or acronym (like VLE) may do. In a less comprehensive form where we have to start, it's a vigorous learning organization (or VLO). This chapter explains vigorous learning organizations and their advantages. Much of it is encapsulated in Figure 5.1, but first the terms in the name must be defined.

Vigorous suggests organizations that do things. Education, as in a school, is not the core mission of most of them, but learning helps them carry out their mission. *Vigorous* suggests purposeful energy—neither a relaxed country club nor mindless, frenetic activity.

Learning is the act, process, or experience of gaining knowledge or skill (dictionary definition). That covers anything from toilet training to

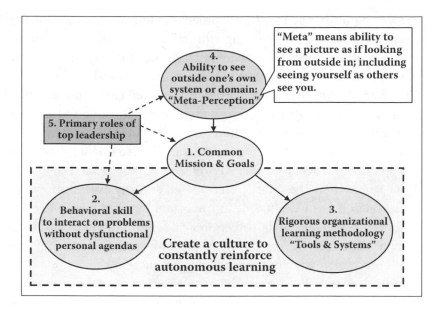

FIGURE 5.1
Overview of a vigorous learning enterprise.

research cosmology. Peter Senge notes that learning consists of at least five disciplines: personal mastery, mental models, shared vision, team learning, and systems thinking. Thus it encompasses the development of individuals, groups, and organizational systems.[3] Chapter 3 concentrates on process learning, innovation, and organizational learning. A vigorous learning organization is intended to pursue learning in all these senses.

Enterprise is used in many senses, but the intent here is analogous to supply chain: several tiers of customer organizations going out and several tiers of suppliers feeding in, plus feeder educational institutions, consultants, advisors, banks, auditors, and the like. *Enterprise* implies relationships with a density of communication greater than those of mere monetary transactions.

Referring to Figure 5.1, the major elements of a vigorous learning enterprise are as follows:

Mission: A mission to benefit society that people actually believe in and work for, and that addresses the challenges of Compression.
Behavioral skill: Patterns of behavior to stimulate and sustain collective learning, but with autonomous action. A box at the bottom of Figure 5.1 says, "vigorous autonomous learning enterprise," a phrase

too long for handy use. Development of skill using a learning discipline constitutes a "learning culture" so contradictory to human nature that it must have customs built into it so that it constantly reinforces itself, or we revert to our normal "political" form.

Common organizational learning discipline, based on scientific logic: This uses a structure or system to record learning and access it. The discipline validates new knowledge against reality and measures whether action is advancing the mission. It also implies developing approaches to deal with wicked problems as they arise.

Meta-perception: The ability to see oneself and any organizational situation as if one were outside it, important for everyone, but critical for core leadership. Meta-perception guides the pursuit of social missions with many factors to consider, and under changing conditions. It's much more than just having financial success.

Servant leadership: To serve the mission and to serve other people, not only core leaders need to be servant leaders, but all core personnel. Everyone has to develop both their technical capability and their behavioral capability to the utmost and become professional.

Autonomous: Minimal command-and-control. Every experienced person is expected to work without detailed direction: improving work processes, creating innovations, and safeguarding the welfare of external stakeholders and natural processes. Autonomous is not synonymous with chaos, everyone doing as they please, but working by the discipline of work standards and a rigorous learning system. That is, people learn to work as teams of professionals.

Vigorous learning enterprises are proposed as revolutionary organizations to deal with Compression on earth; they are not sci-fi imagination. Ideas for them are an amalgamation of some of the best practices of human work organizations today. Real people should be able to work this way, but no precise blueprint is possible. Each organization has a different set of people, different technology, and different problems, so each one must find its own path to becoming a vigorous learning enterprise.

Every little business in the world need not become a vigorous learning organization. It is recommended, *de rigueur*, for organizations responsible for consuming a lot of resources, and on which quality of life depends.

That includes energy companies, utilities, most manufacturers, health institutions, schools, construction, public safety, and similar organizations that now involve up to 40 percent of the total workforce. They have to take the lead, doing what reactive, legalistic regulatory processes can't do—head off many problems before they happen. Ultimately, the public has to deal with Compression, but vigorous learning organizations must create structures to enable them to do it. For example, before the public can recycle effectively, somebody has to organize recycling processes that actually work.

COMMON MISSION AND GOALS

A mission statement states what an organization exists to do, and by implication, what it does not do. Well-crafted, a succinct mission statement should bind everyone together in a common cause, understood and internalized by all. *Internalized* means that they have to reflect on it periodically—make more of it than a few words on a wall. It should clarify an organization's general obligations to customers, other stakeholders, and humanity, while leaving room for judgment calls.

Creating and renewing a mission statement is not a process to be taken lightly. Doing it well resolves one wicked problem (what do we exist to do) in order to preclude many other wicked problems later (can't agree whether *this* is a problem). That is, it defines an arena in which Plan-Do-Check-Act–type logical processes can resolve many problems.

A mission statement concentrates on what we must do, not what we will get. People can unify around what they must do easier than agree on how to share rewards or allocate tax payments. Therefore a mission statement is not a business plan, it's not a budget, and it's not a pro-forma financial projection.

Businesses don't need mission statements if what they do is obvious. For example, jewelry retailers don't need mission statements to decide on product lines and customer segments. However, the challenges of Compression can greatly affect jewelry retailing, and mining does need to be done by vigorous learning organizations.

Diamonds, for example, are prized for their material properties, but expensive because they are so widely dispersed that gathering them takes a great deal of energy. Estimated yields for better grade diamond mines range from 0.2 to 3.0 carats per ton of ore.[4] Processing each ton takes a lot of energy, so a diamond in a display case represents a lot of BTUs that got it there. Decreasing energy yields, increasing environmental safeguards, and social issues have begun to affect the industry. Retail jewelers now have to consider these factors even if they only merchandise diamonds and trade in them. To do this well, they may need to craft mission statements even if they do not become vigorous learning enterprises themselves.[5]

A mission statement is effective only if people are intellectually and emotionally bound to it. Consequently, developing one and keeping it fresh as new challenges arise is a major leadership responsibility. And it has to unify; people will see through a saccharine-coated social mission if it primarily benefits one stakeholder (any of them, not just investors). A mission that no-nonsense people will self-sacrifice for is no trivial issue.

A mission statement should relate to Compression and probably the quality of life. If one can't be drafted, an organization must question whether it really has a role critical to society. Gray areas can be expected. Missions like spiritual support or entertainment may be vital if they don't consume excessive resources. On the other hand, one must question missions centered on, say, energy-hogging space tourism.

A long-term mission much more specific than improving quality of life may be nearly impossible to craft. To make it relevant to immediate pursuits, it has to be embellished by more specific statements of what is needed for quality of life now. An example is the mission statement of Ventana Medical Systems, to "provide innovations in science and medicine that improve the quality of life."[6] However, its current learning is guided by a current goal of that mission, "to find cancer faster." For inspiration, that's hard to top. Subgoals both focus the mission and confer flexibility to refocus it.

Ventana is in a very fast-moving field: instruments and reagents to improve performance of histology laboratories (microscopic tissue examination). But with technical progress, in 10 years Ventana may be into identifying genetic and proteomic precursors of many diseases. If so, a new primary subgoal will be necessary.

Other subgoals may guide organizational or system development to advance the mission. For example, one of Ventana's shorter-term subgoals is talent management—finding and developing willing, able people and developing them to their full abilities. Subgoals or subordinate missions need to be well considered, too. People can't balance more than three or four of these when prioritizing tasks and projects, and then the organization needs to check itself frequently to be sure that everyone is aligned.

Companies today use multiple terms to connote ideas similar to *mission and goals* here. A *vision* usually means some future state or status to aspire to. A *strategy* may be anything from a detailed plan to a general intent. If used with a strong learning discipline, *mission and goals* may produce tighter thinking, and at the same time allow flexibility and initiative in execution.

This not easy. It requires high-energy, high-involvement leadership to align direction and develop people. By contrast, managing by budgets and controls is relatively low-energy leadership.

Working to a social mission obviously differs from working a business strategy for financial results. Business strategies typically have goals supported by budgets, with both monitored to guide execution of the plan. A vigorous learning organization performs first to its mission. Its budgets are either constraints or operational specifications, much like most nonprofit organizations today (including military ones). Neither strict adherence to a detailed budget nor maximizing profit is the overriding objective.

BEHAVIORAL SKILL BUILDING

Behavioral skill for collective learning is acquired. Of all the elements of a vigorous learning organization, this is the most difficult. We behave the way we do because we always did, and because of the environment we are in. If we chronically lie, it is because lying has usually worked for us—got us what we wanted or kept us out of trouble.

Behavior we can observe is what people do or say, so acquiring a different pattern of behavior is learning to do and communicate differently. Those of us who have gone through rigorous military training or rigorous academic training came out behaving a little differently—not totally changed, but behaving little differently. We learned how to behave in that setting. Moralizing about behavior without changing the work environment has limited effect. To change how people behave, one must change what they do. Change what we do and the results we experience from doing it, and behavior changes— slowly—because we all have to confirm that a new behavior works.

Wise leaders of lean and quality initiatives intuitively grasp this. Just to use the tools, one must learn to behave a little differently. If the tools are used to stimulate more learning (seeing problems and taking initiative), behavior changes even more. That's why Toyota leaders *create* TPS; they don't *install* it.

The intent of creating a vigorous learning organization is to move beyond the lessons of Toyota to deal with the bigger challenges of Compression. This takes us into learning in a broader sense as is roughly suggested in Chapter 4 and Figure 5.2. We have to learn more than how to better design and build things to serve a customer.

Change the Environment, Change the Behavior

An old nemesis of company training is sending a few people off for training. When they return to the same work environment, they soon forget whatever was learned. Nothing much changed, including the behavior.

Many managers now understand that dramatic organizational changes require changing the work culture, which is a response to every influence on how work is done: technologies, human resource policies, status systems, IT systems, reward systems, and leadership behaviors. All these amalgamate into "how we do things around here."

In any organization, all practices combine to make it what it is, even if that is utter chaos. Trying to change everything cold turkey is confusing; old practices support the old system, and if new ones are not

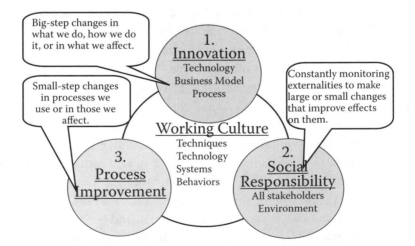

FIGURE 5.2
Three forms of learning.

yet functioning, cold turkey is dead turkey. Just changing a major, organizationwide software package is apt to be described as "organizational menopause." That's why many change programs inch along as continuous improvement: make a change, consolidate and support; make another change, and so on. Because developing a vigorous learning enterprise is a real stretch, its rate of development will be dictated by the speed that people can develop not only new skills, but also new patterns of thinking.

Develop people and their new systems together, and as people develop their understanding of not only what to do but also why, they help develop the system. Each part of a learning culture should reinforce all others.

However, the way TPS developed is worth revisiting. There was no detailed project plan with a budget; only consistent intent to maximize workers' ability to improve their own work processes. To that end, all techniques were trial-fit, then either accepted or rejected into the package. No fixed blueprint with all the answers existed. In time, the system extended into most suppliers (but not dealers) and became codified as the Toyota Way.

The intent for creating a vigorous learning enterprise is learning how to deal with Compression. The scope of an enterprise encompasses more than factories or operational sites. Such a change must displace entire working cultures, not just parts of them, with ones centered on learning—learning the right things to do and learning how to do them flawlessly. This is a leadership challenge that is almost the opposite of blustery, charge-the-ramparts leadership.

A work culture is encoded in symbols, artifacts, stories told, informal networks, and the like. Unless people have a sense of urgency because there is no "normal" to go back to, all this budges very slowly, if at all. Leadership has to create new stories to be told. But even when highly motivated, people cannot change instantaneously.

Think of developing a vigorous learning work culture as a combination of a professional school with military unit training. Only there is no final graduation—no over—ever.

The whole point is preparing for any contingency and for rapid change. This is more like health and military cultures than milking a fat commercial cow. One must be ready for top performance at any time.

But how can we develop people concurrently with their learning system? Start with something seemingly simple, like the four-step training method called Job Instruction (JI) from Training Within Industries (TWI). Almost anybody can learn to train others in any kind of work task using JI. Trainers learn to observe detail more carefully, seeing what they formerly missed. First they must break down the task into steps for learning, identifying key points to emphasize in training. Then they must learn how to coach someone else using the method in their own work breakdown, starting with preparing the task location for training. And no two trainees are alike. One may not consciously understand what they do themselves—or see the flaws in it—until they train others how to do it. Likewise, someone familiar with JI is likely to be a more apt pupil when instructed by someone else—they know what a key point means, for instance.[7] A few formally educated folk may consider JI too elementary, but simplicity is its virtue; it's part of learning to speak the same learning language as everyone else.

A big advantage of training people to train each other is that it *rapidly cascades new skills throughout an organization*. Train some trainers. Soon they have trained more trainers, and so on—a multiplier effect. With this in place, new techniques diffuse rapidly in practice, even with thousands of people. Learning capacity for skills multiplies. But to develop and sustain this capability, people have to practice regularly—maybe even using regularly scheduled learning times.

The same approach can be taken coaching people to use A3 papers or the equivalent, which is more than learning how to use a form; it's a process of learning how to reason and solve problems in a common format so that everyone can communicate. Continuing in this way, engraining the use of a common package of such tools begins to build a learning culture. Any organization can have a package of common tools similar to those used by many others, and it may have unique ones besides, depending on the technology. Some take years to learn, so not everyone can master them all; but the intent is to diffuse expertise as much as possible, increasing the ability of all team members to contribute to learning. In a highly technical, software-laden world, this takes some effort.

But the key is instilling the behavior to learn how to learn, both individually and collectively. Keep going. Don't water down the development. Mentor people to take responsibility for what they do, how they do it, and for its consequences to people far removed who depend on it. Develop everyone until they can improve methods of learning themselves—kaizen their own learning processes. This changes behavior just as military training changes behavior. Lax civilians turn into soldiers—at least in that setting. (Off-base behavior may be very different.)

Vigorous learning in the sense intended here is at least as rigorous as military discipline. It is counterintuitive to normal, instinctive human political behavior that disrupts learning with personal conflict even in academic research organizations. These instincts cannot be eradicated. The best that can be done is to keep them suppressed by creating specific counterpractices and building them into the culture. Leaders being subject to these impulses too, they need to set the example: admit to everyone that they too are human, subject to lapses, and that they expect others to take corrective action with them when it arises.

Collective problem solving is not harmonious passivity, but enthusiastically working a disciplined learning process despite differences. Respectful dialogue (open-minded listening as well as talking) is emotional discipline among people who have done their homework getting facts. Emotional waste is just as real as any other waste, but not as easy to see. Well-meaning people are often unaware that they are creating it.

One countermeasure for emotional waste observed at Ventana Medical Systems is a work culture that Ventana calls a "culture of accountability."[8] That term suggests financial results, and so might be better termed a "culture of responsibility" for a vigorous learning organization in Compression.

Ventana goes beyond most team training today by emphasizing how people behave on the job; half of an individual's pay is determined by how a person did their work, not just what they personally accomplished. Behavior is expected to exemplify the beliefs that characterize Ventana's human work environment. Most employees publicly "sign onto" this system, and are coached in its supporting behaviors. One of those is to speak up if you can't overcome a problem yourself, by saying something or writing it on a board. Of course, co-workers are also expected to promptly follow up and take action, and that, too, is part of the discipline. Another is to keep discussion "above the line," meaning to have an exchange about problems without dragging in personal agendas or personality conflicts. Once into practicing this behavior, a simple word in a meeting is sufficient to keep a participant from straying off course.[9]

Conventions like this in a work culture make emotional discord instantly visible—open to discussion, rather than being repressed and allowed to fester. Such conventions allow people to learn to dialogue—listen carefully as well as speak during discussion. Dialogue capability allows an organization to effectively address problems where emotion can't be excluded. That is, learning to learn is an emotional discipline as well as an intellectual one. The emotional side of learning has to be constantly reinforced to keep human nature from taking over.

A vigorous learning enterprise can accumulate an extensive repertoire of learning tools derived from quality, lean, thermodynamics, and other sources. Software and Web technology expands our access to hyperrational tools; Web 3.0 will push the possibilities even further. However, tools poorly used aren't helpful. Thoughtful carpenters can build more by masterfully using a few tools than tool freaks with a truck full.

Collective learning behavior glues a vigorous learning enterprise together by reinforcing the discipline. Managers addicted to command-and-control may misunderstand this as coddling. It is anything but coddling, like wanting a military unit to be the best it can be if everybody's life depends on it (but not old close-order-drill discipline). It's more like a hospital medical director realizing that the lives of her and her family depend on quality performance by the organization she leads, now and in the future.

Insistence on fact-based learning is as close as a vigorous learning enterprise comes to being a religion. Those who insist on PowerPoint opinion contests have not caught on that this is an anti-BS religion. As they say in quality meetings, "In God we trust; all others bring facts and data."

Fostering the behavior for a learning organization usually has to buck a number of contrary influences: status systems, pay systems, promotion systems, and other trappings of rank and privilege. Despite this, some folks are naturally curious and others are intrinsically interested more in the work than the money. Few may work only for money, but it influences all. However, vigorous organizational learning is for the vigorous, not plodders putting in time, minds on something else. In a nonideal world, creating the behavior for organizational learning has many challenges.

Formal policies of an organization can be altered—with difficulty—but they can't budge attitudes that people carry with them. People have to want to change, and then may remain unaware of what needs changing. For instance, some folks feel compelled to show how smart they are to demonstrate their worth without sharing much insight with others. Some really are geniuses, but in a complex world, even ingenious ideas require input from many others in order to be fulfilled. (It's amazing what works if you don't mind who gets credit for it.) Everyone has to respect other people, but still speak up, so they can take off on the ideas of others and spark collective innovation. People can modify their behavior if they want to, but not everyone is cut out to be a professional in a vigorous learning organization.

On Becoming Professional

Ideally, one would like everyone in a vigorous learning organization to become professional, both in expertise and in attitude. However, the public takes seriously only the oaths made by military personnel or health-care professionals. If the primary purpose of the organization is making money, oaths have minimal significance.

In 1847 the new American Medical Association adopted an updated Hippocratic Oath, the most ancient of all professional codes. It helped transform nineteenth-century physicians from community healers into more science-based learning practitioners. Nurses and other health occupations also began to adopt codes, improve education, and inspire professionals to serve others. At the time, an oath of service was important for the public to accept their professional status.

Companies—and entire industries—now adopt professional and ethical codes. The Illinois Institute of Technology's Center for the Study of Ethics in the Professions has identified at least 850 of them.[10] Professional codes resurface when leaders recognize that something is missing: professional obligations have melted under pressure until conflicts of interest—often monetary—boil up. Commercial performance is rewarded with bonuses, professional performance more with recognition.

Oaths focus on ethics—what you do when no one's watching. Early in the twentieth century, ethical behavior was equated with superior performance, assuming that learning how to perform exceptionally well was an ethical obligation. Elwood E. Rice promoted an Emblem of Business Character for honor, quality, strength, and service, a forerunner of today's more rigorous Baldrige Quality Award.[11] But meaning well does not sustain dedicated professionalism, much less promote vigorous learning.

Free market proponents often criticize professional self-certification as monopolistic credentialing that needs a free-market shakeout.[12] True, most apprenticeships and professional schools cap enrollment. Entry is subject to potential bias, too. But in Compression, professional performance can't be judged only by market criteria. Ethicists criticize professionals for letting subtle lures of money undermine pride in service and quality.[13]

Professionals in attitude really should always be learning and improving in their field. That's what professional practice means. But quality performance to benefit all society is incompatible with constant monetary calculus for self-interest.

Although almost all recognized professionals want more social respect (and pay), not all think deeply about why they might deserve society's respect. Unless fueled by controversy, *professionalism* is defined by snooze-inducing legalisms, something like this: Professionals master a complex body of knowledge and skills based on some area of science or learning, or on the practice of an art founded upon it, used in the service of others. They profess commitment to competence, integrity, altruism, and the promotion of the public good within their domain. These commitments form a social contract between a profession and society, which in return grants the profession a degree of autonomy in practice and within limits, self-regulation. By putting social responsibility above personal reward, members of a profession also assume an obligation to improve both the methods and the outcomes of the processes with which they work.[14]

This wordy proclamation describes living your work, going the extra mile. Dedicated professionals feel a social obligation. They may also truly enjoy it. Financial reward, while not objectionable per se, is secondary. Professional autonomy means self-governance to appraise performance that nonspecialists can't address. However, professionals are not monastically separated from the rest of society. They may flock together, but need social breadth to relate their work to social benefit, and to span the gaps between different technical specialties. In organizations professional specialists may be difficult to integrate. A partial solution is to make everyone in the organization a professional, at least in attitude and commitment to a common mission. This is more than "empowering workers," a radical thought no more than 30 years ago.[15] But if everyone is expected to be professional, they must uphold a social mission that genuinely commands professional commitment.

After 1980, teamwork in American workplaces became popular. However, hordes of employees remain passively cynical—in Dilbert land unable to take much responsibility for their work. Some have professional degrees.[16] Joanne Ciulla describes it well: "Employees haven't been altogether honest with their employers about the way they feel, in part because they are afraid, in part because a growing portion of the workforce has become cynical. Cynics are much harder to work with than revolutionaries because they don't believe in anything; they don't band together into unions; and they don't protest."[17] If sufficiently alienated, even their health suffers.[18] The good aspect of this is that they have enormous potential.

Platitudes are fleeting inspiration. If social need is obvious, as after a natural disaster, people readily pitch in, but day-in and day-out toil in a mechanics laundry, for instance, isn't inspired by blither about quality, efficiency, and environmental sustainability. One cannot expect workers there to become professional degreasers without making a real connection between consuming minimum resources, generating zero toxic releases, and providing long-lived uniforms in which others can make their contribution.

Think of creating a vigorous learning enterprise as creating an all-professional system. In contrast with empowering people, it sets no limit except how much people can develop themselves and their learning processes. Like most ideals, it will never be reached, but going for it builds the kind of organization needed to cope with Compression.

AN INTEGRATED LEARNING SYSTEM

Vigorous learning organizations need to think of themselves as partly being a school, although ground-level formal education would be the mission of only one type of them. In schools today, faculty develop curricula and instructional techniques, but new faculty not in on this redesign them with nodding reference to what did or did not work well before—if they know. The alternative is to adapt a systemwide curriculum to the school setting. Each one is slightly different, and what works for one instructor doesn't for another, so "we're unique" is an easy argument to make. Scholars have researched all this for public schools, but learning it vicariously doesn't help much. Every new teacher has to develop his personal approach on the job. Once settled into patterns that work, he may not be inclined to vary them much. That's work, with risk to it. In short, few teachers have a local learning system by which to improve effectiveness year over year.

This is complicated by the wicked problem of agreeing on effectiveness—on what a school should accomplish. Standardized tests can measure whether students have acquired structured skills, but what about imagination, behavioral attitudes, and much else? We can't resolve this without resolving the wicked problem, so learning in this sense is a prisoner of emotions, too—from school boards to pupils and their parents.

Most work organizations have a similar weakness, even if their core mission is knowledge generation, like university research labs. Their notebooks, databases, software tools, and professional journals are designed to push the leading edge in some field, but much know-how is not standardized and improved unless it seems crucial to results. That is, few have a rigorous learning process to improve their own learning process.

Most companies, agencies, and nonprofits have similar issues, even when they have "lean and quality" programs. These philosophies seldom expand to create an integrated learning system.

"Integrated" suggests that a learning system:

- Covers all the kinds of learning suggested in Figure 5.2
- Isn't just a set of training and software packages
- Is built into how everyone does their work every day
- Enables people to overcome problems wide in scope and deep in complexity

Most companies have knowledge repositories on technology, customers, environmental hazards, service histories of equipment, and so on. These can be expanded and put into self-similar formats so that many people can interpret information "not quite in their area," and the disciplines to use them can be coached. Many companies' information is neither well coded nor well organized for easy access and communication. The A3 format is not the only one possible, and it may not fit everything, but its concept has advantages. It promulgates clarity of thinking, and it begins to simplify the confusion of hundreds of report formats. It also forces writers to think about who might want to find and interpret the information they record in the future, and why.

What kinds of information? Everything from operator instructions at workstations to technical data for "set-based" integration of either produce or process design. (One actually used by a heating and refrigeration company recorded the names of repeat customers' pets—useful for the next tech called to that customer.) Data to support DfX—"design for everything," the environment, test, use, and so on—has to come from outside the company as well as inside. Compiling such open access databases is now in infancy. Perhaps in the future we can even include access to information from Web 3.0 in the learning system. That, for example, might be data from sensors anywhere in the world that we custom-integrate into a map of ship positions in the Pacific Ocean.

The value of these repositories depends on information being entered in them in timely fashion. Some repositories may be well-disciplined Wiki networks, but the key to usefulness is making time to organize entries and enter them in the system. Use of the system is not a passive activity; it's active knowledge sharing, part of regular work, and a discipline. To be effective, time must be allocated for this. In much knowledge work today, time spent keeping up and honing skills is a big fraction of the total, but where the need is less obvious, it is neglected.

Learning repository systems themselves need an improvement process. If they are not used, question and adjust processes until people can't get along without them. And one reason for system disuse may be that people are not engaged in enough process improvement and innovative activity to use it. That, too, is a problem.

Successful use of a learning system depends on a stable cadre of people using it. They may not be locked in as corporate lifers, but high turnover is deadly when complex products and processes must be integrated and well tested, as with twenty-first-century aircraft, vehicles, power systems, communications systems, and yes, recycling systems. Creating a world that can cope with Compression takes more than pickup squads of people who mean well.

Today, organizations still think of hiring talent instead of developing people as a learning team. They train people extensively if the training applies to their operations. They may pay for outside education. But they don't think as much about rapid learning as part of the work itself.

The state of apprenticeship programs illustrates the point. The biggest apprenticeship program is Norfolk-Grumman Apprentice School for shipbuilding, retained because shipbuilding requires unique skills, and development of a work leadership cadre.[19] Many other apprenticeship programs faded because of fear of organizational inbreeding, cost, and the fact that craft apprenticeship training is often done by unions. Instead cost-conscious companies depend on junior colleges for skill building, recruit the best they can find from the job market, and train them as needed. If employee turnover is high, training the influx of new people is costly, but failing to keep them long enough to mold an aggressive learning culture is costlier.

Companies now must be confident of a new technology's reliability before it is released. Long gone is the colorful era when a Bill Lear could palm off car radios as a hot new thing, but that frequently caught fire because they were not grounded.[20] That's unacceptable now, much less in Compression, but many people picture innovation coming from disruptive entrepreneurs like Lear: legendary eccentric, inventor, socialite, womanizer, and temperamental boss.[21] In a century where getting everything right is critical, Bill Lear's rambunctious ways are inadequate.

Instead, if both products and processes must be DfE (designed for the environment)—and for everything else—a great deal of learning must be acquired and ingeniously integrated. Products have to launch a long way down the learning curve. If one-off designs, like molecules customized to remediate one specific patient's malady, are to become routine, fail-safe learning has to climb to a new level. More learning has to be assimilated into useful form faster than ever before. In the code phrasing that describes complexity, the density of learning processes has to increase.

An example of dense learning is the race to use DNA for computation. Such a system is engineered from BioBricks™, snippets of DNA having known characteristics and encoded in an open database. An organization that assimilates the most information about BioBricks™ is more likely to assemble a prototype biocomputer.[22] But snippets of DNA are also used as scaffolding for orientation of carbon nanotubes and for applications, so BioBricks™ have multiple possibilities.

One can think of learning system databases as well-organized information bricks, or better, "live" repositories of constantly updated information bricks about environmental hazards, customer needs, materials properties, product use characteristics, and much else. No enterprise is likely to own all the bricks it needs, so withholding information from others as intellectual property is beginning to be recognized as an impediment to learning.[23] If this truncates the information in databases used to design products and processes for environmental sustainability, it verges on antisocial activity, when what we need in Compression are learning enterprises that collaborate. Addicts to competitive money races don't yet see this conflict.

But none of this creates a vigorous learning organization without people actively learning, creating a different kind of twenty-first-century world. All need to learn above their traditional pay grade. For example, production operators should take equipment apart and reassemble it to learn how it really works. Some of them should even be able to do this with sensitive instruments. And they should learn the software embedded in their work process; otherwise, they are inhibited improving processes. The objective is to be able to deal with any imaginable challenge in a high-tech world in Compression.

METAPERCEPTION

Learning should also expose people to a wider world, so they learn to consider the whys as well as the how-tos of their work. For example, engineers and operations personnel need to see live customers, products in use, and end-of-life disposal or recovery processes.

The sooner this broad understanding begins, the better; but other than occasional bring-a-child-to-work days, children rarely see much of the working world. The adult work they see is day care, entertainment, fast food, retail, and the like. Sewage plants, foundries, hog farms, and operating rooms may as well be on a distant planet. Few adults ever see them either. More can gain an appreciation for natural processes than work processes. Indeed, acquiring more than cursory knowledge of how industrial society physically works takes more time than anyone has, especially when buried in our own grub holes. Industrial society's transactional system obscures much that ought to be seen by many more people. The new system needs to address this.

In the United States, K–12 education preparing people for work in the present system concentrates on math and literacy, plus computer and team skills. In postsecondary education, more people have graduated in business than in any other field, learning its language.[24] However, learning to think broadly and deeply gets short shrift. Scientific reasoning is limited to critical thinking. IT students have at least one course on system integration, and business students have a strategy course. But extensive learning knowledge from many sources must acquire and integrate it. To make vigorous learning enterprise a reality, multiple languages must be learned.

That is necessary to promote change and learning across a broader front than ever before. Preparing people for truly radical innovation is a big plunge into vigorous learning. Rapid process improvement has to become routine. Then invent new business models: reverse logistics for bringbacks; upgrade units in service; lease but don't sell; remotely monitor installed units; extensively observe firsthand how products are used; and of course, research to minimize resource consumption, toxic releases, and festering issues that may appear years later.

Even the definition of *success* isn't the same, so measuring success isn't the same. For example, a popular measure of innovation is the percentage of sales from products new in the last two years. How about the percentage of products in the field that are converted to reduce energy use in the past two years? Or that are remanufacturable, recyclable, or even "upcyclable" (each recycle improves material purity rather than degrades it)? Somebody has actually done each one of these feats now. But way out there, how many of our quality-of-life needs and environmental needs can be met with totally different solutions and technologies? That challenge is beyond a one-idea company.

ROLES OF CORE LEADERSHIP

Top leadership of a vigorous learning enterprise implies the top of a pyramid, which is necessary if it is to carry any authority, but it might be better described as core leadership. It differs from almost any other kind because a vigorous learning enterprise has to behave more as a nonprofit institution than a profit-making company: mission first; professional workers second; then all other stakeholders; financial stability to support the mission; and me last, except for personal development. Enterprise leadership must guide it away from commodity traps that endanger quality of learning and quality of performance. This leadership challenge comes in two parts, external and internal, although they blur together. Leadership of transition is daunting, especially when trying to launch a vigorous learning enterprise, which will always be in some kind of transition.

External Leadership

Migration from a commercial business into a vigorous learning enterprise fundamentally changes its basic premises. The board cannot merely nod approval. Designing a legal charter for full-stakeholder social responsibility is apt to be simpler than getting stockholders to release their grip. The organization needs sources of capital that invest *in* a mission rather than purely for a return, something like high-yield school bonds. There is an obligation to repay those who invested in the mission in proportion to their risk, but not to maximize their monetary wealth. This alone is no small leadership challenge.

Likewise, leadership has to build external support for this change among all stakeholders. Besides customers, employees, and suppliers, the nature of the enterprise needs to be at least understood by auditors, bankers, investors, regulatory bodies, schools, and maybe courts and any other learning partners whose support is critical to success.

Few people are likely to comprehend the mission and nature of a vigorous learning organization from an elevator speech. Many have to be co-opted into understanding. For example, auditors unfamiliar with so much as lean operations may not begin to grasp it unless they personally engage in kaizen to improve a process.

Just to handle external leadership a core leader needs experience and metaperception to sense the perspective of all stakeholders. A leader has to orchestrate their support of pursuing a social mission with stretch goals—without bogging down in never-ending meetings. That is, core leaders have to be persuasive setting direction, explaining it, and sustaining progress toward it.

Leadership: Sustaining Trust

The role of core leaders of a vigorous learning enterprise is like that of music directors, guiding other people's learning, developing them to learn how to play well alone, in groups, and all together—and without a conductor, like improvisational jazz players. They can't pretend to be as expert on every instrument as the players, but they have to know enough to mentor people and to be a role model to follow. The concept is simple: Develop people to collectively learn the real needs of customers and other stakeholders, research technology, and develop processes to serve them extraordinarily well. This development may mature after a few years, but it never stops. No matter how skillful, all musicians need to practice, and most love to play or they wouldn't do it.

Core leaders have to create this kind of work culture; then protect and defend it. To do this, people have to trust them as exemplars of this culture, and their own transformation to do that begins with attitude. One phrase to describe this attitude is "servant leadership," holding the organization and its mission to be more important than personal status. In a business setting, pioneer servant leaders like CEOs Robert Greenleaf or William O'Brien did not pound tables demanding returns to ownership or big rewards for management.[25] Their genuine concern for the welfare of all stakeholders showed. Lip service to servant leadership is worse than useless. In fields like the military, health care, and even politics, servant leadership is expected to sustain a modicum of public trust. If money motives or personal aggrandizement undermine this public trust, charges of corruption fly, with legal indictments in flagrant cases.

In business, this balance is not expected; maximum return to ownership is. If financial secretiveness is employed to fulfill this expectation, trust among stakeholders is easy to destroy, and especially when an organization has to cut back. For example, arcane executive pay packages have been decried by stockholder activists, many of whom prefer that *they* get the money; and if government bailout money is used for payouts to either investors or executives, then everybody protests. Boards and executives bound by fiduciary duty to ownership seem blind to this, unfazed by even cases like Don Carty, CEO of American Airlines, hiding fat executive pay and perk packages from unions while negotiating give-backs from them. Before unions would come back to the table, the AMR board had to dismiss Carty.[26]

In expansion everyone can get a slice of a growing pie, so big differences in rewards are tolerated, but in Compression we can't tolerate this emotional waste. (It's not profitability per se, but attitude that divides; employees like to share gains, but no one likes to share losses, especially if they seem highly inequitable.) Executives' payouts were justified as incentives to align them with ownership, but when sops from owners can't mollify other stakeholders, this idea implodes. When ships are going down, what kind of leaders can everybody trust in the lifeboats?

To align all stakeholders in a vigorous learning enterprise on performance to mission, core leaders have to behave as servant leaders, blunting contentiousness over rewards and recognition (but it never disappears). Then they can unite stakeholders around what they collectively need to do. In a vigorous learning organization, a core leader's behavior has to exemplify that expected of everyone—becoming a lifelong learner. Therefore, a core leader's model behavior has to emphasize lifelong learning, not win/lose negotiating. Once armed with this attitude, a core leader is ready to impart it to others.

Leadership: Developing Others

Leaders at all levels, not just the top, have to shift responsibility to those who do the work as fast as they can handle it. Begin by asking people questions, directing them no more than absolutely necessary. This has two aims. First, coaching people to think more deeply; then, coaching them to take initiative. With practice, leaders learn to ask better questions, provoke people to also question each other and to learn from each other. Good questions correspond to three of the learning loops at the beginning of Chapter 3:

- What have we learned? (First loop)
- What have we done to prevent having to learn it again? (Second loop)
- What have we set up to learn more? (Third loop)

This sequence roughly follows the scientific method, so it starts channeling neophytes' thinking patterns toward more formal problem solving, beginning with pursuing *why* questions to root cause. As a formal learning system is structured, questioning nudges people to use it. This has to persist until the patterns are a daily habit, which may take several years. It may start as a pilot project, but it soon has to spread throughout the organization—from customer service to R&D—until people devise improvements for their own learning processes.

As responsibility transfers to people doing the work, leaders have to relinquish it. This may terrify those sworn to both fiduciary duty and accounting principles, like the stereotypical controller fearful that transactional disciplines will disappear. However, transferring responsibility implies trust-but-verify accounting, not abandonment of all audit trails. Nonprofits need these, too. In addition, anywhere that visibility of work is increased to speed learning, audit trails begin to simplify because simplifying operations (learning) reduces the imaginary complexity that accounting needs to track. Once free from this distraction, people can concentrate more on learning how to better fulfill their mission.

And once having untangled such matters in their own minds, core leaders can concentrate on developing people into professionals dedicated to the mission. Much of this effort is mentoring others in the learning system, although any organization with complex technology and environmental entanglements also has a big load of specifics to keep continuously updated.

Mentor the learning behavior as well as techniques, and mentor some to become learning leaders mentoring others. This starts to cascade the development of people, equipping them for substantive dialogue. Simultaneously, the daily work environment has to adopt disciplined learning structures like standardized work improvement, learning databases, and visibility systems—the syntax of the learning language. Success is when others autonomously assume responsibility for challenges and proudly report what they have done. The most dedicated become professional, always thinking how to better fulfill the social mission even when off-duty.

To change their own behavior as well as that of others, leaders have to stop old routines and mix with the people; the culture won't change if leaders stay closeted. Instigate some new stories for people to tell, stories that inspire people struggling to become something else, displacing their historical lore on "how things used to be." For example, make some dramatically different decisions in their midst, explaining the reasoning, like, "We must refuse to serve customers that refuse to return material to us." Leaders of an organization so big that most people only see them sporadically have to leave a trail of exemplar stories in their wake.

Drive out fear of change. Because leaders are learning along with everyone else, mistakes will happen. Admitting your own errors and openly correcting them encourages everyone to keep learning even when embarrassed. If an idea fails, help others laugh and learn; then try again. If it succeeds, celebrate and go on. It's OK to fail if we learned what doesn't work and don't repeat the mistake. It's not OK to stop learning how to learn better.

As a leader, getting this right is crucial. Button your mouth and let others make their mistakes—as long as they are not catastrophic. Keep your ego in check and resist the temptation to tell others how to solve their problems. It's the only way they learn.

This kind of leadership is high energy, and vigorous learning does stretch everyone, so back off once in a while and have fun. Being constantly frustrated or keyed up is draining.

Some little things are very important for building the trust needed in a vigorous learning organization. For example, at the Naval Academy (which inculcates a version of servant leadership) one of the most arduous lessons learned by plebes is to answer a trivial question (like "What's on the lunch menu?") by admitting "I don't know" instead of trying to bluff their way through. Integrity in such seeming trifles is another base on which trust is built. Minor lapses are forgivable, but when it counts, both military units and vigorous learning organizations must depend on everyone telling it like it is. That element of a learning discipline, once instilled, gives an organization more collective competence and flexibility than command-and-control directives.

There is so much to impart by mentoring that a leader constantly seeks teaching opportunities. The best counseling encounters occur near the time when something has gone amiss, but done after a person counseled is reflective, not angry or upset. Timing this summons every talent a leader can muster. Indeed it is so involved that leaders may need to meet regularly to review what works and what doesn't, coaching each other in how to mentor better.[27]

Leadership: Defending a Learning Culture

At maturity a vigorous learning organization is disciplined by its learning system and using minimal command and control. Although some, like the Orpheus Chamber Orchestra, rotate leadership, all human organizations have a hierarchy. Somebody has to take the lead in setting direction. Someone has to take initiative on nonroutine matters. Somebody has to make decisions when there's no time for confab, as with a ship's captain in a storm. Someone has to represent the organization in public, or talk with irate people or with authorities who want to "see the boss." And someone has to deal with employees who have committed flagrant violations. That is, leadership has to defend as well as develop the kind of organization in which most people can autonomously further the mission with leaders only pointing in a common direction.

In highly autonomous, decentralized organizations, at least four functions are typically centralized: (1) setting of mission and main goals, (2) finance (the bank), (3) human resources (initial selection and assignment of people), and (4) common systems and culture (what makes the organization glue together). However, a hierarchy need not be recognized with formal titles, as at Sun Hydraulics, a leading supplier of engineered hydraulic systems and components in Sarasota, Florida. However, everyone knows the designated go-to person for situations that are outside an autonomous solution.[28]

Internal defense of vigorous learning is as important as external. Not everyone is cut out to fit into a rigorous learning discipline. As it matures, many people who can't cut it will take themselves out; their peers refuse to stop carrying them. An open, visible work environment leaves few holes to hide in. A few may take on roles requiring little interaction, but in a learning environment, those are very few.

The experience of Ventana and other organizations tending toward vigorous learning is that even after careful screening, turnover of new entrants is as high as 10 to 20 percent in the first year. Until they are in it, some folks don't understand what a learning environment demands. A few may regard their exclusion from it as unfair. Fairness of initial opportunity is a necessity, but retention depends on one's ability to keep improving as an individual and collectively. Dropouts can work for less-demanding organizations less critical to quality of life and still be of value to society.

A vigorous learning enterprise has to stress quality performance. Both participation in it and leadership of it must be earned by merit and sustained on merit. The organization exists to elevate human work performance to the highest level it can achieve, so it should not be confused with businesses that exist to maximize income or wealth for owners with minimum effort on their part. There is no coasting.

Therefore preparation for succession of people (or more accurately, for further evolution) is part of upholding the vigor. Here a sports analogy is apt. When developing a championship contender, only the best players, in top condition and melded as a team, are going to be on the field. Vigorous learning is not sports; it's much more serious, but quality of performance cannot be compromised by people with a different objective, thus "improving it worse." If development of people never ceases, there should be no problem with either leadership or followership succession planning.

Finally the leadership must know, or at least sense, whether performance to mission is improving, and whether people are developing. The basics cover five areas: status of organizational learning, customer service, use of resources, financial solvency, and readiness for the unexpected.

PERFORMANCE MEASUREMENT

The performance measures in Figure 5.3 are only some suggestions for an enterprise dedicated to coping with Compression, but they put a little more rubber from ephemeral concepts in contact with the road. Figure 5.3 is long to illustrate several points: (1) a vigorous learning organization

Measurement Category	Example Measurements
	Individual Learning
Status of Learning	No. of group process improvement experiences
	Level of experience with PDCA
	Proficiency in group problem solving
	Qualified to train others
	Direct experience time with "externalities" (list)
	No. and depth of technologies/skills in which qualified
	Capable of autonomous process change
	Recognized internal expert in: (list)
	Can serve as mentor/coach/instructor in: (list)
	Languages in which proficient
	No. of entries in codified learning system
	Leadership development status
	Process Learning
	Life Cycle Innovation
	No. Business process innovations, past year
	No. of new technologies actually deployed
	No. of new technical/natural discoveries
	No. of tech packages in database, for use
	No. of patentable findings
	No. of research exchanges
	No. of tech databases accessible
	Percent of customers served with totally new processes
	Life Cycle Process Improvement
	No. of changes to standard work
	Process visibility ratings
	Percent of processes set up for problem anticipation
	Time to develop new design for deployment
	Time to redesign old product upgrade
	Time to retrofit old field units
	Organizational Learning
	No. entries in codified learning system
	No. of accesses to external data sources
	Time in meetings to resolve issues
	No. of meetings/initiatives to anticipate issues
Customer Service	**Customer Satisfaction**
	Status of customer understanding and acceptance
	Satisfaction with service and service advice
	Renewal of service percentage

FIGURE 5.3

Example performance measures for a vigorous learning organization. (Continued).

	Customer Service Operations
	Time to define customer needs
	Unplanned time without service
	Service response times
	Percent of issues resolved before customer is aware
	Lead time for orders, revisions, upgrades
Use of Resources	BTUs or joules of energy used
	—By products used by customers use
	—By operational processes
	—Percent of energy generated at site of use
	Process yields, defect rates, scrap
(Are we learning to have better	Mass of material processed
outcomes while using fewer	Operational space required
resources?)	Space required to provide field service
	Lead times, inventory levels, ship-on-time
	Distance material moves
	Percent of material reused, recycled, and virgin
	Percent of discard sent to landfill; Percent of biodegradable
	Water consumption
	Effluent: volumes, weights, toxic inventory
	Emissions: volumes, weights, toxicity
	Operational flexibility: variety, substitutions, upgrades
Financial Solvency	Present and projected cash flow position (+ or −)
	Margins from current operations
	Contingent liabilities (risks analyses)
	Capital expenditure conservation
	Credit line, investors, lender, and grantor potential
Readiness	Measure of extent to which mission is being fulfilled
	Alternate "What If" scenarios able to deploy
	Status of safety and disaster training
	Robustness to disruption—flexibility

FIGURE 5.3
(Continued).

differs greatly from an expansionist company, (2) it deals with challenges more complex than most working organizations today, and (3) many of its performance measures are radically different, but not every one of them.

Figure 5.3 assumes an organization that deals with a product line such as food refrigeration with a mission to improve performance for existing customers while greatly reducing the resources used to do so. That assumption makes the list more specific, but any real learning organization has to develop performance measures to fit its own mission and situation.

One big difference measuring business performance is very noticeable. There is no return on investment measure. Financial planning is for solvency, not maximum return. Ability to pay financial backers is important, but not as important as mission. A fast-learning organization using physical footprint measures and anticipating problems will make better use of money than is possible using financial hurdles for decisions.

Not that a vigorous learning organization won't need money; it will, but return on investment doesn't tell anyone whether the organization is meeting its mission. Making money is more like an operational specification; it should not distract an organization dedicated to an important mission. Call this "private socialism" if you wish, but it is what socially responsible companies would do now, were they not beholden to growth demands from the capital markets.

Suppose a vigorous learning enterprise's mission is to use much less energy to fulfill the functions of an existing truck fleet. It sets a goal to cut total energy use of the fleet by two-thirds in 10 years, but alternatives are limited without considering the functions of the fleet. That's why an enterprise is needed. Changes must consider the reasons why customers of a fleet use it.

One line of attack is working with customers (and customers of the customers) to cut the demand for energy: Rethink the fleet's function; reduce the total number of vehicles used and the travel distances of those used. Another line of attack is with vehicles: Deploy radically new-design vehicles, retrofit old ones, and so on. Decreasing the energy use of every vehicle by two-thirds is not the most imaginative way to tackle this. Everything, including all processes intersecting with those of the fleet, is open to change.

Using total fleet energy use as a performance indicator for meeting this goal draws too small a system boundary. The enterprise has to key on energy use of all processes to fulfill the function of the fleet, including the energy yield and environmental performance of the sources of energy to perform those functions, and projected over an extended period of time (as with mass–energy balance and life cycle analysis in Chapter 3). Without drawing a big system boundary and holding to it evaluating performance, it's too easy to outsource some of the functions and pretend that the objective is being accomplished.

A perfect example of pushing off problems is outsourcing American manufacturing to China, where by present standards the production base was less equipped to cope with energy and pollution problems. That move exported the mess from American consumption to other people—classic quantity-over-quality reasoning. By expansionary system measures,

Chinese prospered from this, but they will not put up with its problems much longer. Compression is a global challenge. Coping with it requires global enterprise imagination.

Of course, many other factors are important to the performance of functions and systems now touched by the truck fleet. One is the quality of life of people served by these processes, beginning with such basics as lead-time of delivery and on-time delivery. Then toxic releases and the quality of equipment and building performance (not just trucks) are important: downtime for maintenance, lifetime performance, flexibility for upgrading, potential for remanufacture and recycle, and so on.

Mission performance measures can only be defined after its goals and how to deliver them are in mind. They will evolve as progress is made on the mission; they are not slabs of commandments coming off a mountain. Vigorous learning has to promulgate rapid, detailed, integrated change, not establish new, illusory sources of wealth—cash cows to be milked by privileged people.

To do this, even enterprise business models will change, perhaps no longer selling objects, but a service enabling the functions of the objects to be performed—without hyping turnkey solutions. Everyone has to think. One can imagine no end of uproar over this. Owners are reluctant to kill an old cash cow as long as it gives a trickle, but without dealing with Compression, most cows' days are numbered. New business models have to reward people for using less—the opposite of expansion. An example is PortionPac Chemical Company, Chicago. It packages cleaning supplies in use-size packets, making excess usage difficult, but does not sell the packets. Service contracts, per building or per room cleaned, include educational literature and hands-on training. If it develops customers' cleaning personnel well, PortionPac ships less material and makes more margin on a contract. Customers save money too.[29]

In Figure 5.3, the lead item under Customer Satisfaction is "status of customer understanding and acceptance." Quality of life is primary to all missions, but in many cases customer satisfaction isn't enough. Customers have to change life habits, too, so many vigorous enterprise missions will include a goal of pragmatically coaching customers using appropriate performance measures.

But the most crucial performance measures are of the type that head Figure 5.3, those that attempt to measure the degree to which an organization as a whole is learning how to learn faster and better. Yet, this can't be measured directly, but only by surrogate indicators.

One can also have too many measurement indicators; "data diarrhea" tells no one what they want, while gathering and interpreting the data consumes energy. For central direction, better to rely on a few indicators that everyone can follow. Despite the length of Figure 5.3, once into the thinking, readers can keep adding performance measures to their wish list.

Indicators also need to represent something consistently. This is a problem today when data are captured differently depending on how semantic slush is interpreted. For example, "on-time shipping" may mean anything from "arrive when the customer wanted," to "ready to ship but not actually sent." Aggregate roll-ups of performance measures have to be eyed dubiously because of interpretive differences in data capture, or differences in actual operations measured the same way.

Next issue: Who uses performance indicators to control what? Present-day managers often want to use a dashboard of a few KPIs (key performance indicators) to drive an organization as if it were a machine. Without a strong learning discipline, what they are actually driving is more like a herd of cats, but they go for the illusion of top-down control, with the KPIs substituting for financial controls.

Budget compliance and other control measures to hit a profit plan are conspicuously absent from Figure 5.3. In a transactional economy, cash planning is essential for vigorous learning organizations, too, but profit is not its mission. Except governments, no organizations can long spend more money than they take in, but responsible people can actually exercise better financial judgment without detailed budgets.[30] Professionals autonomously minimizing use of resources perform better without them.

COMPRESSING YOUR LEARNING PROCESSES

Viewed from an expansionary business mind-set, all of this is strange. To help clear cobwebs, some key distinctions of a vigorous learning enterprise are:

- Mission, not wealth maximization, is their primary purpose.
- They are composed of people. They do not work to primarily benefit owners. The people need a focused professional commitment to serve all stakeholders.

- They must become capable of quickly executing well-considered changes. That is the key to coping with all other challenges.

Naturally, proposing vigorous learning enterprises as a pragmatic approach to twenty-first-century challenges stirs apprehension. They are a venture into the unknown that deviates substantially from prior history. Although nothing suggests that they are impossible, they are a major step toward increasing the human disciplines that accompany an advance in civilization. They will be opposed by people fearful of letting go of the past no matter how much evidence screams that we cannot continue on our present path.

Vigorous learning enterprises move beyond normal free-enterprise work systems. In practice, these may be great, or be subverted by nepotism, cronyism, bullying, and payoffs (the negative side of almost any human system). We can no longer abide in the incompetence of undisciplined systems, whether free enterprise or state socialism, but our biggest challenge lies within us: overcoming the human weaknesses—egotism, parent/child mind-sets, money status—that are highly divisive. We are a long way from developing the capacity for collective learning and collective action that characterize advancement of human achievement.

No textbook prescription can create vigorous learning organizations. The details and specifics of each case must be dealt with by people experiencing them firsthand, thinking as they go. This book can do little more than suggest a general framework, for there is no such thing as The Answer, a holy grail that if found will make everyone live happily ever after.

ENDNOTES

Extended version of endnotes available at http://www.productivitypress.com/compression/footnotes.pdf.

1. An example is *Low Carbon Energy: A Roadmap*, Worldwatch Report 178, 2008.
2. Ronald Wright, *Short History of Progress*, House of Anansi Press, Toronto, 2004. "Progress trap" refers to societies, ancient and modern, overrunning their resources too late to easily retrench.
3. The disciplines are briefly defined at http://www.solonline.org/organizational_overview.
4. Figures on carats per ton are from Mining Watch Canada.
5. Diamond industry initiatives on these issues are at http://diamondfacts.org.

6. Robert W. Hall, "Culture of Accountability: Ventana Medical Systems," *Target*, Issue 6, 2007.
7. Jim Huntzinger, "Why Standard Work Is Not Standard: Training within Industries Provides an Answer," *Target*, Issue 4, 2006. See also http://www.twisummit.com.
8. Robert W. Hall, "Culture of Accountability," op.cit.
9. "Above the line" is from Partners in Leadership, LLC (http://www.ozprinciple.com). ABOVE THE LINE®, BELOW THE LINE®, and STEPS TO ACCOUNTABILITY® are its registered trademarks.
10. A listing of these codes can be found at the IIT center's site: http://ethics.iit.edu/codes/index.html.
11. Elwood E. Rice, LL.D., *A Tribute to Business Character*, Rice Leaders of the World Association, New York, 1929.
12. James Fallows, *More Like Us: Making American Great Again*, Houghton-Mifflin, Boston, 1989.
13. William M. Sullivan, *Work and Integrity*, HarperCollins, New York, 1995.
14. The definition is an extended version of the *Oxford English Dictionary* definition.
15. M. Scott Myers, *Every Employee a Manager*, third edition, University Associates, Inc., San Diego, 1991.
16. From Scott Adams's comic strip. The author, when visiting offices, has often counted cartoons on the conjecture that trust in management is inversely proportional to the number of Dilbert cartoons posted per cubicle.
17. Joanne B. Ciulla, *The Working Life: The Promise and Betrayal of Modern Work*, Three Rivers Press (Random House), New York, 2000, p. 164.
18. Michael Marmot, *The Status Syndrome*, Times Books (Henry Holt), New York, 2004, p. 122.
19. Northrop-Grumman Apprentice School at Newport News, VA, is probably the best-known American program.
20. Harry Petrakis, *The Founder's Touch: The Life of Paul Galvin of Motorola*, Motorola University Press, Schaumburg, IL, 1991. Several stories agree with Bill Lear's accounts—heard personally from Lear.
21. Richard Rashke, *Stormy Genius: The Life of Aviation's Maverick Bill Lear*, Houghton-Mifflin, New York, 1985.
22. One can start learning about this at The BioBricks Foundation: http://bbf.openware.org.
23. For example, Roche's policy to share information can be read at http://www.roche.com/sci-genepatenting.pdf. Gene patenting is not a simple issue to resolve.
24. See http://www.census.gov/population/www/socdemo/education/sipp2004w2.html. (Table 5A).
25. Servant leadership in business is most closely associated with Robert K. Greenleaf, *Servant Leadership*, The Greenleaf Center, Indianapolis, IN, 1977.
26. Wendy Zellner, "What Was Don Carty Thinking?" *Business Week*, Apr. 24, 2003.
27. Robert W. Hall, "Tokyo Seksui," *Target*, Issue 2, 2008.
28. Sun Hydraulics's policies are at http://www.sunhydraulics.com/cmsnet/aboutus.aspx?lang_id=1.
29. PortionPac began with "green chemistry" in 1964. See http://www.portionpaccorp.com.
30. Robert W. Hall, "Eliminating the Budget: Park Nicollet Health Services," *Target*, Issue 5, 2007.

6

Developing the Constitution for Vigorous Learning

We haven't begun to come to grips with the challenges of Compression. As long as our imaginations and our systems are still gripped by expansion, we can't become serious dealing with them, but only tweak what we have been doing.

To muscle up to these challenges, we need a global constitution whereby we can agree on what to do. Devising anything pragmatic that resembles such a thing requires humanity to learn faster than ever before, and emotionally as well as logically. This book proposes a way to start.

Dictionary definitions of *constitution* disclose several meanings:

- Setting up or composing something
- Health or strength of a person
- Fundamental rules or principles prescribing a government or other institution

As used here, *constitution* implies all of those, plus a stronger meaning: agreement on how to agree.

The global financial implosion of 2008 is a minor tremor. The earthquake has begun where people have already exceeded the carrying capacity of their regional environment, as in East Africa. There, many people respond by tribal instinct for self-preservation: Winners live and maybe even prosper; losers die. That is what a global constitution must avoid. To survive, some people in East Africa may need to move elsewhere; to have a better life, almost all need to learn how to *do* differently. However, all of us need to learn how to *do* differently, which is what a global constitution must promote.

In the complacency of an expansionary environment, the challenges of Compression are hard to get in mind, then keep in mind. No one can track all the threats to human physical system expansion; Chapter 1 could present only a few, and undoubtedly many remain unknown by anyone. We have created resource shortages ourselves. Many pushback threats are rooted in the excesses of pursuing them. Directing our array of technology on these challenges, from nanotech to intelligent software, requires super-competent organization. That's one reason for proposing vigorous learning enterprises as the action organizations to deal with these challenges.

One reason we are complacent is that our psychological health depends on optimism—faith that no matter what, we will survive, assuming that "if half the world's population were wiped out, I would be in the surviving half." A second is that the economic logic of expansion—our macro and micro tools for decision making—assumes that expansion will continue because for several hundred years, it always has. We are now so dependent on these tools that figuring out how to do anything using different logic and tools will be a major learning experience.

DEEP CHALLENGES

Given our present human population density, restoring an imagined eco-logically pristine earth is impossible. Any scenario to drastically reduce the human population conjures images of genocides past no matter how idyllic the motivation. Turning back the technological clock—luddite reasoning—might doom many of us, and every superhero technical miracle has a downside. No option is easy or obvious. We have to meet the challenges of Compression with the maximum performance humans can muster.

Every human consumes resources and puts CO_2 in the air just breathing. However, industrial societies consume just to be consuming—for status and to provide jobs—while those living on the margins barely stay alive. But everywhere humans waste far too much and foul their own nests. For example, no one knows how much food that is grown never enters a human mouth, but is spoiled or thrown away, and some that later makes people sick. Malnutrition (by overeating as well as undereating) is only one problem we face. We have to do much better.

One example of doing better at doing good is the Hunger Project, one of many nongovernmental organizations that help people in developing countries learn how to help themselves.[1] The key is to find or to develop a community change agent that comes from the local culture, but has been outside it long enough to envision how it could improve, and may have skill in health care, teaching, and so on. To be effective, this vision and the will to pursue it have to be seized by the local leadership and populace: There is a better way; they have to find it. If literally starving, people are receptive to change. Begin by creating a community center for health care, training, and community meetings—a place to explore ideas outside the customs that froze them in their status quo, but everybody has to learn their way. Food production and preparation can often be increased just by relieving women of care for children so they can work the fields. Could the Hunger Project do better? Of course, but people only progress at human-change speed.

The Hunger Project seems a world away from higher-tech projects to reduce the energy required by the functions of a vehicle fleet, or a Ventana eager to help pioneer new routes to quality of life based on the genetic "diseasome." It really is not. We are all reluctant to give up old ideas that "worked for us" as sources of wealth, status, or acclaim. Even scientists fear that investigators of a different paradigm will threaten their grants.

Coping with Compression has begun in various ways, but it has to focus and accelerate, becoming a worldwide movement. Of course, policies by various governments or international bodies are necessary, but rapid change occurs when motivated people are able to do it directly, not manipulated into it.

And what we have to do globally is momentous:

- Cut the use of energy and virgin raw material; cut wasteful consumption.
- Eliminate the release of toxic chemicals and take care introducing new molecules into the environment.
- Increase the ability of people everywhere to not only survive but work toward a better quality of life.

That is, we need to globally create at least the same quality of life as in industrial societies today, while using less than half the energy and virgin raw materials, and cutting toxic releases to nearly zero. Quality over quantity, always.

TOWARD A STRONGER CONSTITUTION

Constitutions of democracies generally provide for enacting and enforcing laws that all citizens will formally accept (if not always abide by), but craft vague compromises in areas in which people disagree. Citizens can never totally agree on definitions of marriage or fair systems of taxation, but most will accept majority rule so long as it does not violate some deeply held belief (their tribal religion).

Democracies allowing input from everyone can exist as long as people have big areas of agreement and tolerance for their disagreements. Where this is impossible, one group establishes an empire, imposing rules by various means of control, without pressing those ruled to the point of revolt. Empires generally unify people around some glorious ideal that counteracts tribal loyalties—free enterprise; rule for the proletariat, and in its least imaginative form, the mystic "divine right of kings." People sustain empires as long as they improve their lives—create a higher energy yield to them than when existing as separate tribes.

Among investors, corporate governance is a form of dollar democracy, but to other stakeholders, it is an empire. Avowing the purpose of a corporation to be income for investors simplifies its intent, but not its processes for technological development, customer satisfaction, or operational execution. And it doesn't deal with Compression.

Now a global economy hinging on corporations has become a transactional empire with its own system rules. This resembles a global economic constitution that isn't very effective, but it affects all of us, even the pedal cab drivers of Asia. Like it or not, all governments that participate in the global economy, including socialist ones, are subject to it, too. The global capital markets are the central province of that empire, blowing and popping equity market bubbles, concentrating wealth and dissipating it. It giddily guided growth when resources were to be had aplenty, but now that guidance system is rudderless. Yo-yo pricing offers few signals to cope with Compression.

We've only toyed with modifying this constitution. Some financial funds invest only in socially responsible organizations. A few boards draw employees into more responsibility through gain sharing, ESOPs, and pension plans. However, these initiatives hardly touch the main system. To deal with Compression we need a much stronger constitution to promote much greater density and speed of learning.

Given the global crisis of financial confidence, global conferences will impose tighter rules on monetary systems and markets. These may prop up the system of expansion for another round or two, but a stronger global constitution needs to evolve, encompassing more than transactional rules and standards of measurement. It has to accomplish at least three things:

- Dampen tribalism to make more "wicked problems" possible to resolve, and promote the ability of people to identify with a global "human tribe of the whole."
- Promote a process theory of value; stop making so many decisions based on money calculus that promotes expansion (quantity), and start making decisions that cope with Compression and promote quality.
- Recognize a system of audits, not just to prevent decay of the disciplines needed, but to help organizations learn to adopt them faster.

Establishing this constitution as a single, written document is impossible. Who would meet to agree? Could they foresee what needs to be done more than a few decades into the future? Could they follow through with anything they agreed to?

Instead, this constitution has to be about learning—learning how to agree about much more, much faster. Begin bottom up, where people can deal with specifics—with vigorous learning organizations. With each move, try to reduce residual wicked problem areas, where apparent conflicts of interest seem to make issues irresolvable. We could afford the old system when we could recover from almost any kind of error, but now our safety cushions are shrinking. If the civilization of a space sovereignty (Chapter 4) were no more competent than ours, its inhabitants would quickly perish.

Bottom-up developmental learning is the principle behind vigorous learning organizations. Although these may seem farfetched, they begin to deal with specifics at ground level. That seems much less farfetched than trying to agree on a global constitution first. Vigorous learning organizations have to simplify complicated systems just to make them manageable and learn how to overcome human issues. In due course, perhaps they can develop much of the human population to the level of understanding that lets a global constitution for learning "evolve."

WHY THIS APPROACH?

Four systems of human work organization are shown in Figure 6.1, which attempts to graphically summarize them. Many of us are still trying to rise from Tribalism at the bottom of that figure. One can certainly see examples of tribalism in industrial society organizations today.

Before getting to a Strong Constitution for Learning at the A-level of the diagram, an alternate hope for humanity is Transhuman Technology at the B-level.

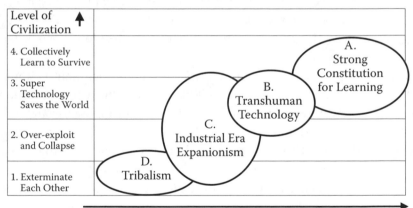

System	Implications for Civility and Social Discipline
A. Strong constitution for learning	Rigorous organizational learning discipline, differently motivated. Extensive development of holistic human capabilities required.
B. Transhuman technology	Great confidence in technical solutions; limited anticipation of undesirable outcomes; little change in expansionist economic thinking.
C. Industrial era expansionism	Shows signs of incompetence. Slow boom-bust learning; systemic biases for growth; complex transactions; can easily regress into tribal destruction.
D. Tribalism	For much of the world's population this is the "natural" state; win-lose systems whereby winners take scarce resources from losers.

FIGURE 6.1
Advancing our level of civilization.

Transhumanism envisions a future dominated by enhanced humans: better bodies, better minds, and better lives. It supports development and use of technologies like nanotechnology to do this. Forerunners of this are all around. Drug enhancement is an old issue in sports (including horse racing). Night vision, artificial reality, and prosthetics are not new; entertainment themes based on bionic people came and went, and we have long had planes that can't fly and cars that won't run without computers. Moore's law is still holding, and we are experimenting with quantum computers that could be capable of solving problems that humans can't even conceptualize. Transhumanists hold that these trends are in infancy.[2]

One of the best-known transhumanists is Ray Kurzweil, who contends that computer intelligence will eventually exceed that of humans.[3] If so, software has to learn to improve itself, but *why* would software want to keep improving itself: Just to prove it can? Self-preservation? Preserve others, too? Maximize profit (what's profit to a software package)? The questions presume that software can self-motivate, that is, adopt human-like consciousness and question the purpose of its existence. But if it can, superintelligent software would face paradoxes similar to those of humans: Am I becoming smarter just for the sake of being smarter—existing just to exist? Do I seek a better quality of existence? If so, what is that, and is it just for me or for others, too? Does "others" include humans? If so, all of them or just some of them?

These are not idle questions. Humans trap themselves in their own dangers because they avoid addressing such questions, or when they do, they can't come up with answers acceptable to all parties. That's one reason we band in tribes. And transhumanism illustrates the paradox of empty technology—knowledge without a purpose (although one may appear later), or its purpose generates conflict (as with stem cell research). For example, we may soon discover how to medically adjust human brain uptake of dopamine, dampening urges to take dangerous risks—use more logic and less emotion. But suppose extreme risk-takers want to keep taking them, and that some businesses catering to high-risk behavior risk failure if they stop? Transhumanism advocates ethical development of technology—ethical from a human viewpoint, not a computer's, so it must eventually deal with the same issues as Compression.

A strong constitution for learning has to create more ways to agree and resolve conflict-laden issues. Compression and a constitution for learning

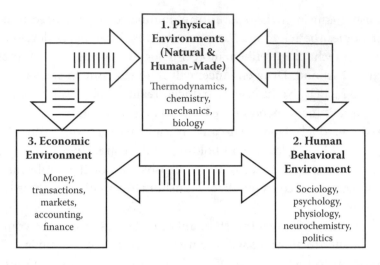

FIGURE 6.2
Three major domains of thought.

assume a purpose: Assure survival of *all* humans and improve their quality of life on a fragile, resource-short planet.

So this constitution has to expand thinking beyond some economic framework. It has to consider the three domains of thought in Figure 6.2 and maybe more. These roughly correspond to the three forms of learning in Chapter 5, and to the triple bottom line popular with environmentally conscious companies (people, planet, profit).

I have been amazed by the gulfs of misunderstanding separating intelligent people mesmerized by one of these three domains. Some seem incapable of accepting that issues in another domain so much as exist. Although many financial executives are very broad thinkers, this gap is significant for people trapped in finance-think, with their rewards based on it, oblivious to the human dynamic within their own organizations, to how its physical operations actually work, and to environmental issues. The most extreme seem to think that enough money can change the laws of physics. On the other hand, critics of business may harp on its failings with little understanding of financial systems, customer needs, or physical constraints, imagining that magic would happen if executives merely curbed their greed. Crossing these divides takes more than cursory learning. Shouted debate, rather than genuine dialogue, only widens them. To start closing these gaps, perhaps acquaintance with biology and thermodynamics should begin in the second grade. People cannot exercise reasoned judgment serving society with a professional attitude alone.

A GLOBAL TRIBE

Tribalism, not our expansionist legacies, may be our greatest impediment coping with Compression. To Westerners, the term *tribe* conjures small non-Western ethnic groups, perhaps like the Yanomamö of old (see Chapter 3); Western tribes are *clans* or *mafias*. The Palestinian-Israeli conflict is a modern version of tribal dispute.

More generally, *tribe* is an epithet occasionally used to describe groups sharing deep convictions about politics, religion, economics, or work roles. Departments in a company may behave as functional tribes although the people in them are ethnically diverse.

International news regularly features ethnic conflict, as in Iraq, Rwanda, or Gaza, but to distant observers unaware of their histories and myths, the causes are incomprehensible. Violence erupts, outsiders act as intermediaries, and violence abates; but hatred lingers. In the grip of old revenge cycles, upholding honor and justice quickly reverts people to savagery. Melting this away to build trust can take generations.

Historically, tribal conflict subsided if a dominant tribe imposed an empire, or if everyone agreed to live by a constitution—a framework of rules for collective decisions and conflict resolution. But tribalism lurks below the surface of a constitution's veneer. If tribes lose confidence in its processes, tribal conflict breaks out, often as street riots in industrial countries; for example, Paris (North African immigrant youth, 2005) and Los Angeles (Rodney King arrest plus blacks vs. Korean shop owners, 1992). Social and economic discrimination often contribute as factors.

Tribalism shapes how we frame problems. If a strong ethnocentric bias is absolutist, tribalists think that if everyone else thought as we do, all problems would be solved. By contrast, the constitutional ideal is common allegiance to a set of principles whereby diverse people can accept—if grudgingly—decisions made procedurally. Trials and debates seek evidence and weigh it. In practice this has holes; else no one would be wrongfully convicted, but it beats tribal lynchings. And as with the

U.S. Constitution, tribes try to bend it to their own beliefs, overtly and covertly. It's called politics.

Rational attacks may, for example, point out impracticalities in one-size-fits-all laws and regulations. But because tribal bias frames all issues and questions from its own belief system, tribes cannot agree on how to agree. Taboos may even prohibit tribes from seeking facts about differences. Tribes can coexist only if these are never discussed. Tribalism justifies win/lose mind-sets, making many problems too wicked to resolve. The roots of tribalism probably reside in how the brain learns as described in Chapter 3. However, tribal thinking can be shifted by reframing problems using propaganda. George Lakoff has summarized the science around this in popular form.[4]

Implications of Tribalism at Work

Tribalism pervades twenty-first-century work organizations, too. Market systems are a step up from no-holds-barred, but self-interest thrives. Company competition, departmental budgets, and bonuses stimulate competition between "tribes" playing by the rules. Successful entrepreneurs bulling through them to win are often compared with athletic competitors. Governments and standard-setting bodies set most of the rules. Companies usually abide by them—but lobby to change the game to be fairer to them. Civil constitutions create ways to mediate (or referee) between competitors. Money making is a contentious game with rules bloated by the minutiae of precedents from many seasons of contention.

This generally worked in expansion, but if markets shrink, companies desperate to survive financially (to win) are tempted to skirt rules regardless of consequences to others or to the environment. At worst, they revert to mafialike tribalism, every family muscling in for themselves, like a feudal system or despotic economies today.[5] Zimbabwe is a current example of a market system reverting to feudalism.

Laissez-faire economics stripped of Adam Smith's moral sentiments about them amount to little more than mafia rules; hence, never-ending arguments over how much market regulation is enough. In practice, socialist alternatives easily degenerate into tribes headed by political operatives instead of managers. In expansion, much of the disputation is about fairness among stakeholders. Systems to improve fairness among stakeholders range from neocorporate negotiations to the Mondragón cooperative in Spain.[6]

However in Compression, more effective, quality work systems are a top-priority need, so we need to reframe the ideological disputes hanging around from expansion. Fairness among stakeholders is just one more wicked problem for a global learning constitution to resolve.

Tribal instincts are a pervasive impediment to vigorous learning. Regarding all humanity as members of the same tribe is contrary to our tribal nature, but necessary because vigorous learning enterprises depend on trust—trust in the professional responsibility of others to seek and abide by facts and data, wherever they may be, regardless of ethnic tribal culture. Having no trust beyond the tribal confines impedes learning.

Tribes seldom welcome negative feedback. Loyalty to the group—or fear of it—trumps seeking or accepting facts. Those seriously questioning a process are ostracized or degraded as whistleblowers. Leaders don't have to banish them. Loyal subjects do it for them. Where *omertà* (mafia code of silence) prevails, forget becoming a vigorous learning enterprise constantly questioning purposes and processes.

This behavior still infects work organizations of the twenty-first century, regardless of form of ownership: government, nongovernment, and for-profit. Two well-known whistleblowers are Jeffrey Wigand, who ratted on the tobacco industry (as told in the movie *The Insider*), and Sherron Watkins, the Enron vice president who wrote the red flag letter to CEO Kenneth Lay.[7] Most whistleblowers are at a minimum shunned internally, reprimanded, and warned not to speak out. If they go public, their employer will not only fire them, but also try to discredit anything they might say or do, blacklisting them to all society, especially the media. Many are even disowned by their families, consigned to a lonely existence in a land of the living dead. Few would do it again.

For instance, Peter James Atherton in 1975 reported to the Nuclear Regulatory Commission a hazard with electrical control cables and other flaws at the Maine Yankee Nuclear Power Plant. Ignored all the way up the commission's chain of command, he finally tried to see President Jimmy Carter, was handcuffed, committed to psychiatric care, and fired. But he was right: After a short circuit started hydrogen fires in 1996, Maine Yankee shut for good. Today Atherton, vindicated but uncompensated, still lives alone in a Spartan apartment on income from odd jobs.[8]

C. Fred Alford, an expert on these cases, concludes that rage at whistleblowers' insubordination buries all facts, corrective action, or ethics in malevolence. Right or wrong, whistleblowers rarely anticipate the price they will pay. Most sacrifice their careers and social status for their professional principles, or for attempting to do the right thing. But postwhistle life is so miserable that support groups have formed to assist them, for example, National Whistleblowers Center at http://www.whistleblower.org.[9]

Of course, whistleblower claims do not always accord with facts. But their travails point out a crucial weakness when work organizations do not seek indications that processes need correcting and immediately follow up. Any vigorous learning organization has to overcome instinctive human resistance to openness and abiding by facts because we fear that we will get ourselves or our group in trouble—if not with a boss, then with somebody (like a customer company) that holds a money whip over us. Leadership has to dispel this fear.

Trust and Tribal Bonding

As noted in Chapter 3, human bond density appears to develop naturally if people get to know each other in multiple settings—the tribe. If separated by 15,000 kilometers, face-to-face contact is impossible. Computer links partly compensate, especially if body language can be read, but time still limits interaction to a finite number of people. This difficulty may be why market rules presume little more trust than necessary for trading—and then with auditing safeguards. When even that breaks down, industrial society begins to break down.

Dealing with the complex problems of Compression will take much more information density than market theory presumes. Complex, software-laden products cannot be developed, provided, or serviced by parties using no more communication bandwidth than cost/price/quantity. Think of it as a high-density market—free interchange of ideas and information with a lot more trust in motives.

Expansion evokes a land rush mentality, with self-interest duplicity in abundance. Compression evokes more of a lifeboat mentality. Does everyone on the boat belong to the same tribe—the human tribe—and try to survive together, or fear to sleep lest they be thrown overboard? A one-tribe feeling isn't trivial, but redefines social status, leadership, work, and success. Accepting it emotionally is gut wrenching, but not all learning is intellectual.

The need to increase trust is urgent. For example, attempts to corral the subprime mortgage debacle revealed that if trust is broken, central banks working in concert are not enough. No infusion of liquidity can reprime trading in markets when traders don't trust either their value or the balance sheets of other traders. In addition, this mess revealed that when expansion goes in reverse, fallout hits almost everybody. Whether shrinkage is big financial bubbles popping, resource shortages, or ecological limitations, getting it right is in everyone's interest—a social obligation. That is, there is no such thing as totally unfettered private property, free of any obligation to anyone else.

The trust necessary to deal with Compression is deeper than that which must be restored to global credit markets. Financial backers of a vigorous learning enterprise must invest in its social mission and quality of performance, not economic growth. They have to support the enterprise's mission and know how vigorously it is meeting it. Few banks have ever been challenged to shepherd organizations learning ingenious ways to limit global consumption of energy, water, and virgin materials—and making a profit by helping customers buy less. If an enterprise's mission is vital, but market revenues are insufficient to support it, could any institution other than government be the financier? For example, suppose almost the entire vehicle fleet of the United States has to be capitalized and operated as a leasing service.

Modest trust is needed for financing. Deep trust is needed among people working in linked organizations to integrate advanced technology and all other considerations needed to cope with Compression. To develop trust, avenues other than normal bonding experiences may be open, maybe even the modification of brain chemistry.[10] However, embracing the mission as a deep personal commitment makes a big difference. Once we accept that crucial work organizations need professional performance both in competence and emotion, many divisive legacy issues melt away. How a work organization is financed becomes almost irrelevant if there are no important distinctions between public and private property, personal and social obligation, economic and social progress, and economic and environmental progress. But by reinforcing these distinctions, our legacies reinforce many tribal beliefs dividing us.

The instincts that assured tribal survival in a hostile world assure mutual destruction in a global, interconnected world. Every step up the ladder of civility in Figure 6.1 requires greater human trust. Break that trust, and civil society at any level tumbles back down to the bottom rung. That happens from maneuvering to win, playing PR disinformation games, not following through with agreements as promised—business behaviors that aren't unusual. If a street-fighter mentality dominates, nobody trusts any statement or promise, so there's no point in discussion. We revert to Yanomamö shamans sensing the evil intent in each other. And humanity perishes of its own incompetence.

A PROCESS THEORY OF VALUE

A process theory of value is unique because it is nonmonetary. The contention is that many of our most important decisions should not be based exclusively on systems dependent on human valuation. Both the market theory of value and the labor theory of value do.

Problems with Existing Theories of Value

Market theories of value generally trace to Adam Smith's seminal work, *The Wealth of Nations*, written in 1776, before the industrial revolution got rolling. It was a guide for developing the British Empire: increase productivity in England, expand free markets to obtain raw materials from elsewhere, and enjoy the wealth of the world. Smith contended that expanding markets increased the revenue from division of labor. His pin factory example improved productivity mostly by task specialization, but his concepts fit nicely with substituting fuel energy for labor and creating mass markets. It was a prescription for expansion.

However, Smith realized that markets don't solve everything. Being familiar with agriculture at the time, he recognized that it required experienced judgment balancing many factors in greatly varying conditions. For good yields year after year, the capital stock had to be conserved: fields left fallow for a year or two, a fourth or so of the ground reserved to feed work animals, the best of a herd kept as breeding stock, and so on.[11] If well-managed, a farm sold for a higher price, so Smith considered expert farmers superior to skilled craftsmen. Expert farmers needed multiple seasons of learning, plus the ability to devise imaginative do-it-yourself countermeasures for unexpected problems. That is,

Smith described an eighteenth-century version of process thinking or systems thinking.

Agronomics, a branch of agricultural economics, continued research and teaching in this way for decades, balancing three factors of production— land, labor, and capital (the stock of animals and equipment required)— and sometimes a fourth, management or methods. Early twentieth-century agronomists researched actual farm practice to maximize its year-on-year yield in the long term. That's a form of environmental sustainability. Back when much that is marketed today was done at home, university home economics did the same. But agricultural research splintered into specialized science fields. Mass-market demand drove the industrialization of agriculture to become a network of contractors, dramatically increasing output, but making old holistic approaches become passé.[12]

A simplistic market theory of value is that a thing is worth whatever buyers and sellers agree to exchange it for at the time.[13] By contrast Karl Marx emphasized a labor theory of value: that everything ultimately results from human work, so anything sold for a value exceeding the intrinsic value of the labor to provide it cheated the worker.[14] Smith's version of the labor theory is a twist on the market theory: a worker will value something in proportion to the amount of labor she will exchange for it as a buyer, but that from the seller's view, a price is the sum of labor, capital, and profit. This difference symbolized the Communist-capitalist ideological splits of the twentieth century. However, both systems tried to maximize industrial and agricultural output, and both depended on some kind of accounting.

Modern market theory pegs value by exchanging items priced in a currency used as a common reference. With growth, exchanges became more and more abstract from direct trading of physical objects. Wealth came to be the present value of a future stream of income: mortgages, loans, credit cards, or stock in a corporation. This "money now for money later" mentality expects a return on monetized investment proportional to perceived risk. The percentage of GDP represented by financial services in the United States came to exceed that of manufacturing and grew several times larger than that of agriculture. How much bloat is in this, exchanging just to be exchanging, is unknowable, but is suggested by statistical shifts.[15]

Psychologically, we buy and sell objects. We invest and divest capital objects. In abstract form, these objects are only representations of value, like certificates, or 0–1 strings in a computer file. Abstract representations are mentally objectified into digital quantities for trading, and financial minds can monetize almost anything. As these monetized representations stray further from any supporting physical reality, their market value is only confidence that someone somewhere will pay a future stream of income. If someone's likelihood of repayment becomes dubious, these valuations enter risk-management limbo where payers and their circumstances are difficult to unwind.

Objectification for trading is the core of capitalist valuation. It motivates physical activity. Money reserves for investment are built by trading at a profit, or created in advance by collateralizing the appraised values of inert things into tradable markers. This magic converts things like coal in the ground into cash, motivating people to mine the coal. Cash thrown off from mining is invested in further ventures or funneled into health care, pensions, and other transfer payments. The system is symbiotic with physical expansion, the combining of human and fuel energy to create industrial society. But if excess money chases unattainable real growth, it blows and pops one dysfunctional bubble after another.

Although growth economics is based on transactional valuation, people hold many values not derived from exchanges. Marriage is one. In any culture, if sexual interchange is strictly for money, it is usually called something else. So how do people determine value?

Legions of philosophers and theologians have probed this indepth, but to simplify transactional models, economists often assumed that value was determined by exchanges between rational people maximizing their self-interest.[16] Over time this caricature came to be called "economic man." Unfortunately, economic man is difficult to distinguish from psychologists' definition of a sociopath.[17]

In the latter twentieth century did behavioral economists began weaning the economics mainstream from economic-man rationality, but its assumptions continue to influence business behavior.[18] For example, the efficient market hypothesis holds that new information disseminates in markets so quickly that buyers and sellers are equally informed.[19] This assumption also props up *caveat emptor* arguments for failing to satisfy customers. For instance, when selling software or insurance, marketers can persuade confused buyers to trust them on faith, but if faced with lawsuits later, argue that buyers fully understood. Reality is that only those who daily review complex contracts like insurance understand every clause, so the exchange is really based on trust.

A market theory of value depends on transactional valuation. Its flaws are illustrated by ancient adages comparing the value of a drop of water in the desert with a drop in a flood. Prices are supposed to cover the costs of efficiently providing a good or service, but when social values are loaded on them, they distort. We impose sin taxes on liquor and alcohol, but subsidize milk and sugar. And the prices of oil and gas, which are now commodities as vital as water and subject to the same valuation extremes, also became obvious weapons in wealth transfer wars and political bludgeoning—another reason for price volatility. "Tribal" political pressures easily suppress dialogue about this, which is why dealing with Compression is such a big human problem.

Market signals mislead when the assumptions of a market fail, for example, that a higher price will increase supply. When this cannot be done because global supply is limited, a conventional market ceases to exist. We're stuck with some allocation method, fair or unfair, controlled by a monopoly or by some other form of governance.

Market-based valuation also has limits of absurdity, epitomized by a 1997 project to value the entire world ecology.[20] A series of estimates averaged $33 trillion with a range of $16 to $54 trillion. At the time, global GNP was $18 trillion. Because all life depends on the global ecology, analysts knew that no value based on dollar exchange had meaning. They went ahead because the importance of the global ecology could only impress many executives (and economists) if it was assigned a huge dollar value.

At the heart of this is a great philosophical divide. On one side are those who regard concern for nature as a do-gooder's drag on the global economy in general and their company's finances in particular. On the other side are those who see that their consumption is just one more digestive burp in a global ecological sickness—their company and all that it affects is codependent on a nature that does not negotiate. Those who cross this divide can begin to discover how to cope with Compression.

However, despite its flaws, an industrial economy cannot function without transactions. Market signals enable billions of actions daily without our having to think much about them. A process theory of value has to preempt the market theory of value for important decisions when we must try to understand their implications on global society and global ecology. Many natural processes (and some human ones) are like stealth bombers—unseen on radar screens—but these screens detect only monetary values, until something blows up.

This will not be easy. None of us can be aware of all the processes that keep us alive and provide quality of life. Just to begin trying to understand them, we need to balance our thinking among the domains shown in Figure 6.2. Nature's language will not translate into money; we have to learn its language.

But at present, money is not only the language of business, but much civic discourse in government as well. To learn this language, people get an MBA. Because the syntax of this language is transactions, process relationships and process changes translate awkwardly into it. For example, lean accounting struggles just to express in cash flow formats the financial benefits of eliminating obvious physical waste from industrial process. Likewise, scientists can't explain their work to legislators or others who frame all issues in the narrow context of costs, prices, jobs, and economic growth.

One reason this mind-set is hard to shake is the false precision of monetary valuations. Numbers to two decimal places lend undue credibility to fuzzy or incomplete assumptions. Intangibles (in accounting speak) are not easily monetized. An example is organizations' learning and innovation processes, although they may preclude missteps having litigious consequences—or others even more drastic. Sequestering ideas as intellectual property seems natural to this competitive mind-set, but even well-intentioned marketing of intellectual capital can stir great opposition, like Monsanto's patented genetic seeds.[21]

Valuing natural processes is wholly outside this system's framework. It can't value something that it cannot monetize in some way as a transaction, so it struggles to put values on a newborn child, disasters that didn't happen, or a true nature preserve (where very few humans ever tread).

Attempts to Overcome Weaknesses

The weaknesses of the present business system are well known, and modifications have been proposed to bolster it. A little-known one is complexity economics, which uses transactional measurements but factors process feedback into cost/price/demand relationships using thermodynamic and biological analogies.[22] Balanced scorecards and corporate social responsibility (CSR) reports have gained more traction; both may be used for internal guidance of a company as well as for external reporting.

The balanced scorecard is the more structured of these two. It usually has four perspectives: (1) financial, (2) customer, (3) internal processes, and (4) learning and growth. It guides strategy across a broad front, deploying it by cascading performance measurements from top-level objectives to detailed ones. Roll-out with employee feedback may be called *hoshin kanri.*

On the other hand, CSR seems to mean whatever a company thinks it means. Wikipedia alone lists eight standards by which companies can report social performance. Googling CSR turns up everything from tooting charitable contributions to a strict measurement of performance according to Elkington's triple bottom line (people, planet, profit). CorporateRegister.com lists many CSR reports and picks an annual winner. Its 2008 winner was Vodafone, a cell phone company.[23] Not all CSR reports are fluff; those like Sekisui Chemical Company are too detailed for that.[24]

The *Economist* argues that CSR is often misguided, little more than self-congratulatory greenwash. However, few big companies can now afford to ignore it. The public and at least 10 percent of all professional investment managers now expect more from companies than mere financial results, so the CSR movement is picking up speed.[25]

Critics of nonfinancial reporting note that with no solid supporting theory, it is a mishmash that lacks measurements by which companies can easily be compared. That's true; this thinking is still in gestation and comparative ratings are not the only consideration. But even in its current state, CSR starts revealing process effects that a transactional framework misses. Critics also argue that the number of interrelated factors to consider is potentially infinite. That's also true, but it is the point of CSR. Taking off the transactional blinders opens a new world of serious concerns.

Balanced scorecards and CSR move us in the right direction, but they graft concern for some of the challenges of Compression into an expansionary mind-set. To vigorously deal with Compression we have to address many more concerns. A few are raised by a cursory review of Figure 6.3.[26]

From indicators in Figure 6.3 one can estimate that American vehicle fuel economy rose from about 12 mpg in 1970 to about 17 mpg in 2005. However, the total number of vehicles grew by 220 percent, so vehicle fuel consumption rose 88 percent. Between 1970 and 2005, America's total energy use per capita stayed almost constant. However, population grew 45 percent, so total energy use grew by about the same amount.

	1970	1980	1990	2000	2005
Population (thousands)	205,052	227,726	250,132	282,430	296,940
Total energy used (quads)	67.84	78.28	84.73	98.98	99.89
Housing (thousands)	69,778	87,739	106,283	119,628	123,925
Reg. vehicles (thousands)	108,814	155,796	188,798	221,475	241,194
Vehicle fuel (billion gallons)	92.3	115.0	130.8	162.5	174.3
Miles of road (thousands)	3,730	3,955	3,880	3,951	4,012
Vehicle miles traveled (billions)	1,110	1,527	2,144	2,747	2,989
GDP (chained to year 2000)	$3,772 B	$5,162 B	$7,113 B	$9,817 B	$11,049 B
Dow-Jones Stock Index	800	838	2753	11,501	10,783

FIGURE 6.3
Physical growth versus financial growth, United States, 1970–2005.

Compared with other industrial societies, in 2005, America's energy use per capita was 231 percent greater than Japan's and 58 percent greater than Australia's, which unlike Japan also has a geographically distributed population. American GDP-to-energy use ratio almost doubled between 1970 and 2005, beating Japan's GDP-to-energy rise of only 55 percent, but Japan started from a much lower energy base and its GDP growth rate was double America's (data from the Institute of Energy and Economic Statistics, Japan).[27]

Crowing about who beat whom using such statistics is useless. First, total energy consumption in both countries (and most others) is still going up. Just slowing physical growth does not deal with Compression. Second, no one knows how much transactional fluff is in GDP, although GDP per capita is a surrogate indicator of quality of life if incomes are not so skewed that bottom feeders starve. Third, compensating for currency variations that distort such comparisons has a limit. Morals of the story: (1) Pay attention to actual physical activity (like energy consumption), not how "we" are better than "them." (2) In making important decisions, pay more attention to physical consequences, perhaps one or two process steps removed, including cumulative growth that sneaks up on us (like the rise in vehicle population).

What a Process Theory of Value Must Do

Market theories of value accentuate the positive—what we can get—with minimal thought for the morrow. A process theory of value has to address shadowy negatives impossible to monetize, and do so from a broader perspective.

For example, the value of an egregious medical error is monetized only by subsequent litigation; lesser ones create waste difficult to capture with transactions. And the value of preparation for a hurricane can only be monetized later, if at all (but insurance companies try). However, the industrial world now depends on processes like electrical power generation that are easily taken for granted until they fail. And these are small issues compared with the challenges of Compression.

A more apt descriptor may be a theory of process evaluation. Many processes vital to us cannot be monetized and assigned a market value. The best we can do is assign priorities to various natural and man-made processes that indicate their importance to human welfare. Which ones *must* we pay attention to, and which ones can we pay attention to only if we have time? Doing this requires anticipating and evaluating systemic relationships and their consequences, not the value of items.

For example, an organizational learning system is an intangible, but assessing it is very important, even if partly subjective. We're forced to use surrogates like: (1) inputs to the organization's codified learning database, and inquiries from it; (2) length of time to come to agreement (shorter is better); (3) percent of time spent in remedial activities; (4) number and scope of breakthrough innovations; (5) suggestions made and implemented; or (6) percentage of people up to date in "what-if" emergency drills (somewhat like pilot training). Other surrogate measures can be devised to approximate the number, quality, and significance of learning cycles.

This theory also helps size a learning organization. While a hierarchy is necessary for some purposes, a vigorous learning enterprise is foremost a problem-solving network. It's not effective if humans can't work out problems quickly. To do that, they need time to absorb, react, adapt, and transmit (another variant of the PDCA cycle). This bandwidth is hard to measure, but it's analogous to optimizing neural traffic in the brain (see Chapter 3). Too big, and it spends too much time communicating with itself, wasting free energy. Too small, and not enough is connected to accomplish much. Design (or create) it for high information density—quick to access and quick to interpret.

To be efficient under changing circumstances, work processes must adapt quickly. Such intangibles are approximated with indicators such as:

- Lead times
- Changeover times
- Transport distances
- Error rates, error consequences, and remediation waste
- Readiness percentages (on-call availability, quality capability, etc.)

But perhaps the most important one is time to reorganize, ready to do something different.

Vigorous enterprise learning is something people *do*. No amount of money can buy it. Investment is necessary to develop people to their full capabilities, but they will do that only if they really want to.

Therefore, a vigorous learning enterprise can't be bought and sold. Only assets, the things people work with, can be bought and sold. Valuing humans on a balance sheet, which is sometimes suggested, implies that they are property—slaves.

Short-sighted business managers reason that there is no point in developing people to a professional level; they might demand raises or quit. Most managers have better judgment, but a process theory of value should spring them from such traps buried in the market theory of value.

Process Evaluation

A process theory of value needs a methodology very different from financial reasoning. Process evaluation is itself a process—a logical process to examine and reason about a processes' history, or chains of causality, leading to its present state, then projecting how it might change other processes. Real people do this. An example is a physician's doing a diagnosis, which often cannot be done in-depth without referring to colleagues.

The world consists of natural and man-influenced processes ranging from nanoscale-sized to the entire biosphere—or the cosmos. Process evaluation must measure effects of human-influenced processes on the biosphere, which is a near-infinite set of interrelated processes.[28] These evaluations are inside out; we examine the effects of our products and processes on the external environment. All interrelations and effects can never be understood, but we are remiss not to probe them as far as we can go. While process evaluation applies to all stakeholders, for most practitioners the environment is the most different one, so examples will emphasize it.

A process may not be assigned a single numerical value, as in monetization, although evaluating it may require considerable quantification. But overall, we can assign a process a priority rating depending on how important it is to human survival and well-being.

Process evaluation: Rate natural and man-influenced processes' significance to humans. Survival is top priority. Quality of life is second priority.

The objective is to characterize processes based on many factors: scope, resource footprint, toxic impact, persistent effects, and interactions with other processes. Any practical system has to evolve from many people's experience. The scale below is just for illustration:

Essential (ES)
Beneficial (BE)
Undesirable (UN)
Hazardous (HA)
Unknown; can't rate (UK)
Energy intensive (E)

Trying to assess processes right now is a jumble. Figure 6.4 is not proposed as an ideal; much better will be devised if many people start working on it, but what is needed is a method to summarize what is and isn't known to alert people to start digging deeper—you could be affecting something really important.

If one has the data, three numbers can be assigned to a process: cumulative energy consumed, carbon footprint (usually carbon dioxide equivalent of greenhouse gases), and ecological footprint (biologically active land area needed to support a process). Several studies suggest that mankind has already exceeded its available ecological footprint, but they are not precise. We have a lot to learn to become really competent at this.

For example, the EPA classifies hazardous *materials*, not processes. However, materials are hazardous because of the processes in which they are used—how they are made, stored, used, and disposed of. If permanently sealed in a container, none pose a threat. But each material could be used in hundreds of processes, and toxic potential depends on quantity used and potential interactions—not always easy to project. For example, pyrite (FeS_2) in coal is a precursor to acid rain, but if separated from coal and reburied where it cannot oxidize, that does not happen. However, the processes for handling and storing coal ash do have risks, because if spread around, a number of chemicals can contaminate both soil and water.

Testing thousands of chemicals for toxicity to biological life is exhausting. Capacity for "wet lab" testing all possibilities is limited, and the number of potentially toxic substances keeps accumulating, so the EPA predicts a chemical's toxicity from its molecular structure using a program called ToxCast™. Work organizations must rely on open databases from the EPA and elsewhere. And unfortunately, toxicity is just one factor in process evaluation.

Only five example processes are shown in Figure 6.4. The first, plankton blooms, is a natural process on which we depend for oxygen, and that underpins the food chains for all oceanic life—not something to mess up, even if we do not fully understand it. Genetically modified corn has obvious benefits, but also many unknowns about long-term consequences. Chromium plating has well-known hazards. Most *Escherichia coli* seem to be beneficial to humans, but *E. coli* 0157 causes human illness, and we don't know enough about it.

The last process example, archaea life cycles, is just beginning to be understood, therefore rated *unknown* in Figure 6.4. Archaea are anaerobic, high-temperature microorganisms deep in the earth, with total mass great enough to affect the ecosystem.

Associated Process Ratings (Feeders and Followers)	Example Processes	Overall Rating*
ES, BE, BE, UK, UK, UK	Plankton bloom	ES/ES/UK
ES, BE, UN, HA, UK	Growing genetically modified corn	ES /HA/UK
BE, HA, UN, UN, HA, UN, UK, E	Decorative chromium plating	UN/HA/ UK/E
HA, HA, UN, HA, UK	*E. coli* 0157 propagation	UK/HA/UK
UK, UK, UK	Archaea life cycles	UK

* Format of Overall Rating:
  (Overall Rating) / (Worst-Rated Associated Process) / (Significant Unknowns) / (Energy Intensive)

FIGURE 6.4
Proposed process rating system.

A process map and rating for chromium plating is shown in Figure 6.5, but not in great detail. Chromium plating is filled with deceptive hazards. None is likely to kill you instantly, so a few people even do chromium plating at home. (You can buy a do-it-yourself kit at http://www.caswell plating.com/kits/plugnplate.htm.) "Decorative" chromium plating yields a shiny surface for cosmetic purposes. By contrast, "hard" chromium plating is for engineering purposes: chemical resistance, wear resistance, or lubricity that is seldom easy to obtain by other means, a distinction important for weighing benefits.

Chrome is a generic term that includes substitutes for real chromium plating (like Alcoa's Dura-Bright® aluminum surface treatment). However, when seen side by side, people usually prefer real chromium. There's no escaping value judgment. The process was awarded a UN rating as undesirable because, given all its drawbacks (HA- and UN-associated process ratings), decorative chromium plating is not necessary for quality of life. Even the BE rating on fume exhaust is dubious because it is unnecessary if plating isn't done at all. Ratings are from a strict interpretation of the definition of Compression. The global auto industry long ago designed out most of the decorative chromium plating that adorned vehicles thirty years ago.

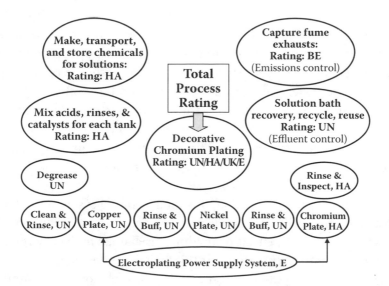

FIGURE 6.5
Decorative chromium plating process evaluation.

Although popularly called metal plating, this industry is better termed metal finishing because a quality finish requires more than plating. For durability, decorative chromium plating requires at least a nickel plate substrate, and probably copper as well. That's at least three dips in different plating tanks, plus lots of cleaning and buffing. Although the process map in Figure 6.4 is skimpy, hexavalent chromium (a carcinogen) in the plating bath is only the most notable of hazards. Hard chromium plating for engineering uses can increasingly be applied by thermal spray using a less toxic high-velocity oxygen fuel technique, but it doesn't produce the shiny surface of decorative chromium plating.

Metal finishing is not an energy hog like electro-furnaces, but it is energy intensive. The electroplating power supply system (rectifying AC to DC current) was rated UN because it drew power from the grid presuming a nonsustainable fuel source for generation, but plating current is just one draw on energy. "Electroless" nickel plating draws no current for the plating step, but energy is needed to heat tanks, degrease, rinse, move material, recover metal ions from solvents, and filter the fumes. Improved process design can decrease energy use, but not eliminate it.

Figure 6.5 maps only processes inside a metal finishing operation, not those of supplier operations, customer operations, or recovery and reuse operations. The Make-Transport subprocess in the upper left oval was rated HA because some storage processes were known to be hazardous, but actually most source processes and transport processes were unknown (UK). To be complete, an overall rating should cover all processes in a life cycle. Today this is not simple, and it is often impossible. We don't know where things came from, how they were made, or where they go. Lead in toys from China is a perfect example. To perform a complete product life-cycle process rating, one has to examine all subprocesses in a full life cycle of a product, whether they are cradle-to-cradle or ore-to-landfill.

As used by a specific operation, this kind of rating system would begin with a mass–energy balance (see Figure 3.6), and then extend into qualitative examination of the effects of anything discharged to the environment. This requires more extensive process thinking than those stuck in a market theory of value are apt to digest at first. It is also obvious that companies organized as now are incapable of this. They have to open up and share learning with each other. Governmental agencies such as the EPA, FDA,

OSHA, or even DOE are even less trusted because managers fear that they will add cost burdens. Enforcement by bureaucracy is wasteful at best, and compliance is even more onerous when one doesn't really want to. This system of self-reinforcing incompetence has to change. And all of this goes back to the issues of egos, power games, tribalism, trust, and fiduciary duty.

But trust and idealism are not enough. People who want to make headway dealing with Compression need to reach out—help each other learn what to do. Information is very fragmented. For example, a manufacturer may want to redesign for the environment but have no idea of the energy requirements or environmental impacts of parts and materials selected. Dr. Paul Chalmer of the National Center for Manufacturing Sciences (NCMS) is compiling databases that will be open, and that will serve as a a generic start point—just a starting point. Significantly, one of the bigger issues in this is proprietary information; companies don't want to openly disclose their exact processes for competitive reasons. The NCMS is a neutral third party, and so databases will not identify specific sources. There is a long way to go here.

NEW CONSIDERATIONS

If we all lived on a patch of ground, grew everything we ate using hand labor, lived in a hut, and never traveled, we would consume much less. Industrial societies have now accumulated so much that they can improve its quality; some undeveloped areas still need the basics to survive. In total, it's quite a challenge to give everyone the means for a good quality of physical life without global overconsumption. To do it we need to pay attention to considerations that we have usually neglected.

New Consideration: Accumulations

Accumulations build up over time. A well-publicized example is polychlorinated biphenyl (PCB, a carcinogen) accumulating in fat tissue of fish. Regularly eating PCB-laden fish causes the PCBs to accumulate slowly in human fat tissue. Industrial production of PCBs phased out in 1977, but PCB levels will remain hazardous for decades because they concentrate right on up the food chain, and they degrade slowly.[29]

Accumulations and slow declines are both sneaky. Slow-building ones like PCBs escape attention until something noticeable triggers an alert—which may not happen until a tipping point is disastrous. Chlorofluorocarbons catalyzing the ozone hole were a sleeper. Fishery collapses have been repeated many times over. Peak oil is another case of denial delaying action. Little changes building up are shrugged off.

Everyone on earth is affected by the cumulative effects and subsequent imbalances from fossil fuels, one of nature's densest accumulations of stored solar energy. Calling petroleum "buried sunshine," Jeff Dukes's back-of-the-envelope stab illustrating this became a well-publicized jaw dropper: A gallon of gasoline is derived from ninety-eight tons of ancient live biomass. Fossil fuel consumption in 1997 equated to harvesting 22 percent of all land plants growing on earth that year. All the fossil fuel burnt since 1751 roughly equates to 13,300 years of plant growth over the entire planet—about a 52-to-1 compression of solar energy. No wonder it can't be sustained.[30]

Industrial societies have accumulated a great deal of infrastructure, put in place by expenditure of great energy as well as labor. If we are entering a lower energy-yield future, the objective is to turn this quantity accumulation into a quality accumulation.

New Consideration: Life-Cycle Energy

Private vehicles make good examples. Suppose a person now uses 1000 gallons of petrol to drive 20,000 miles per year (20 mpg). After process revisions, suppose she goes 20,000 miles on 250 gallons—80 miles per gallon—a stretch, but not beyond the laws of physics. Sounds good, but this comparison just looks at liquid going through the fuel tank.

Suppose the fuel that got 80 mpg comes from a source with only a 1.25-to-1 energy yield (burn four gallons to get five, netting one). Suppose the fuel used to get 20 miles per gallon comes from a supply process with a 10-to-1 energy yield (good by today's standards). Assume that yields factor in energy losses for transport, and that the fuels going into the tank have the same energy density. Then the nominal 80 miles-per-gallon performance was really 16 miles per gallon, while the gas hog got slightly over 18 miles per gallon.

One cannot compare fuels without evaluating fuels both from "drilling the hole" to emissions from the vehicle. Ordinarily we don't have all the information we'd like. Crude oil from various sources may be mixed; so may refined fuels at any stage all the way to the vehicle engine. That's why such analysis is rarely done, but relying on comparisons for which measurement is easy, like the mileage once the fuel is in the tank, is not good enough. Another factor is fuel energy density; diesel has about 10 percent more energy per gallon than gasoline, for instance. It's easy to be fooled.

Take another example comparing two vehicles: one gasoline powered, the other all-electric. To make it easy, assume that 1 BTU of energy applied to the wheels moves each vehicle the same distance, one inch (can't go far on just 1 BTU). What has to be done to get that 1 BTU to the wheels?

To make it simple, suppose that the total well-to-vehicle process for the gasoline car burns 6 BTUs to put 5 BTUs in the tank, again neglecting complications of different fractionations from refining and blending.[31] Of the 5 BTUs going into the tank, only 1 BTU makes it to the wheels (highly efficient by today's standards); radiated heat, exhaust, and mechanical friction claims the rest, so we burned 6 BTUs to move one inch.

Suppose the electric car's battery is recharged from a coal-fired plant. A lot of energy is lost to mine coal, transport it, convert it to steam, and scrub out pollutants. This process will do well if it burns 6 BTUs to put 2 BTUs through the transmission lines to the plug charging the vehicle battery. From there, say that rectifier losses (charging the battery), battery inefficiency, line losses and heat and mechanical losses in the vehicle itself burn up one more BTU. So in total we burn 6 BTUs to apply 1 BTU to the wheels.

Neither vehicle is a clear choice based on energy efficiency alone. What about pollution? For starters, coal has about half the energy density of gasoline. Much of that loss is ash: fly ash up the stack (about 80 percent) and bottom ash in the combustion chamber (about 20 percent). At present, ash usually has to be beneficiated to convert it into by-products like wallboard and synthetic lumber. Only about 30 percent is good enough quality for this, so 70 percent of it is parked somewhere—landfills, old strip mines. Ash and unburned coal mine tailings contain trace metals (copper, arsenic) that cause environmental damage if they leach away from the storage site. Heating and moving this ash makes coal burn less efficiently than oil, so deriving the same energy from it lofts more carbon dioxide into the air. Not all the sulfur can be removed, causing acid rain, which has been a well-known hazard for decades; and trace mercury from coal is contended to be another health hazard.[32]

In addition, high-capacity batteries for electric cars also must be recycled. Recycling of batteries has just begun to receive media attention, but it is one reason that energy storage for alternative vehicles looks at everything from compressed air to supercapacitors.

Remediation of pollution from internal combustion engines has steadily tightened since the introduction of catalytic converters in 1975. Now recycling of platinum, palladium, or rhodium used as catalysts is becoming so profitable that in some places, thievery of catalytic converters is a problem. Nitrous oxide (NOx) adsorbers (not absorbers, but adsorbers) are starting to phase in. All of these measures also sap energy efficiency.

Had ethanol proponents looked at energy yield before launching campaigns for marginal yield processes, they would probably have saved everyone a lot of trouble, especially when corn for ethanol is grown on cropland that may be needed for food. Energy decisions are too important to be made just on claims of dollar potential or job promises. Without this, debate is mostly a lot of sound-bite noise and expensive lobbying.

This seems complicated before getting into it enough to start trimming the imaginary waste from it. The simplest solution is to just use less of everything. Any mess not made need not be cleaned up.

New Consideration: Life-Cycle Operations

In Compression, use of virgin materials must also be minimized. If we need them, reuse them or recycle them. The possibilities for minimizing the use of virgin raw materials in construction have barely begun to be explored. Operations to do this systematically are in infancy, but the Web already has many sites about it.[33]

But vehicles capture the imagination. Germany led the automotive industry toward reuse and recycling by imposing take-back laws. A vehicle cannot be legally scrapped; the manufacturer has to take it back. In the 1990s, German auto companies began designing for disassembly and coding materials for recycling. Disassembly centers were set up throughout the land. As of 2004, the system worked, but only about 20 percent of the vehicles deregistered in Germany were recycled.[34] The rest were presumably exported as used vehicles. Take-back designs did not seem to hurt market competitiveness of new vehicles, and it was short of revolutionary. Personal experience is that German auto companies remain focused primarily on new vehicle market competition—expansionary business, with take-back a side issue.

Remediation of an existing global vehicle fleet of nearly one billion vehicles to improve fuel economy and recover materials hasn't been seriously contemplated by German automakers, and probably no others, and yet technology exists to do a great deal—like optimize combustion in each cylinder over a vehicle's lifetime. The industry is still focused on new vehicles, where new ideas have always hit the market. An industry capable of upgrading a billion vehicles now on the road has yet to form, so back channels for remanufacturing and recycling are minor businesses using a basically different business model. Experience of companies now in remanufacturing is that control of items returned (called *cores*) is the key. If something has been used until it is junk, the best you can do is recycle it. Instead, upgrade it while it is still possible to do.

Is this possible? Quite likely. But it is not a "bubba business." It will take the reorganization and concentration of the brainpower now in the industry on such a mission, or in other words, conversion into vigorous learning enterprises.

Without knowing any thermodynamic parameters, in general, the lowest energy path to reconstituting the use of any item, from a dishpan to a skyscraper, in descending order, is as follows:

- Reuse as much material as possible with no configuration change.
- Reconfigure or refabricate previously used material.
- Recycle previously used material, usually by remelting or by chemical transformation.
- Extract and refine virgin raw material.

The biggest obstacles to innovating in this direction are our legacies from expansion—its operating processes, its marketing mentality, its financial conventions. Shifting from sell-and-forget to life-cycle service and responsibility is no small transformation for either businesses or consumers. The business models are different and the economics are different.

New Consideration: Thermodynamic Currency

To deal with Compression, we need more than awareness that everything depends on energy or fleeting news burps about tons of biomass per gallon of gas. Too few people use Howard Odum's concept of embedded energy, or "emergy," the accumulated energy that has transformed a material or system into its current state. A system may be a small bit of material or a large ecological zone. Energy stored in mass for a long time can be released from it quickly, as by combustion, or slowly through degradation. Awareness of emergy and cumulative effects helps us understand how burning fossil fuels can unbalance the atmosphere—and the ecosphere.

Plant photosynthesis stores a small fraction of solar BTUs in biological cells. Dead cells form hydrocarbons if they sink and slowly compact under pressure over geological lengths of time, thus compressing many seasons of solar energy in high-density fossil fuels. Burning them releases a lengthy accumulation of solar energy, while burning recent growth uses energy at a rate closer to that at which ecological cycles replenish energy from the sun. The excess carbon dioxide pumped into the air from burning hydrocarbons is less than 5 percent of the volume of carbon dioxide that nature generates otherwise, but when this persists, along with other greenhouse gases, the cumulative effect is enough to help shift the balance in atmospheric composition that is associated with global climate change. After the excess burning stops, a prior climatic state will never restore. Countervailing forces will slowly reestablish ecological energy and chemical equilibrium in a new, hard-to-predict climatic state. In short, we humans have to anticipate climatic changes and adapt.

Overall, the ecosphere is not energy efficient; it's a symbiotic dance of life exchanging energy. To survive, organisms become extraordinarily efficient at capturing wisps of energy in every niche of the ecology. Humans are the only species capable of employing vastly more energy than necessary to sustain bodily life, so we need to know more about what we are actually doing when we use it.

However, few people are conscious of thermodynamic balances, much less analyze them. Only specialists use mass–energy balances or Odum's emergy. Most of us look only at fuel economy that we can translate into monetary cost covering only part of a system. A practical problem is obtaining data, but despite the obscurity of these methods now, they are necessary to estimate the effects of human processes on the ecological system, both locally and globally. It's better to crudely measure something important than to ignore it.

Think of energy as nature's currency, thermodynamic currency. Energy cannot be seen directly, so ecologists translate its effects into rough analogues, like ecological footprints, the land area to support a person or activity while sustaining a solar energy balance. By that measure, the ecological footprint needed to support the entire earth's population at the American consumption level has been estimated to be roughly two earths.[35]

Energy balances invariably show that rural areas exporting food, raw materials, and fuels lose net energy (emergy) to importing areas (urban). Preindustrial empires could not create such big energy imbalances. They could not amass the quantities of material objects as modern nations, regions, and companies today, but the quality of what they did do is frequently marveled when seen in museums and old ruins.

Energy, including embedded energy, is nature's currency—thermodynamic dollars that are good anywhere in the universe, independent of human values, and not subject to inflation or deflation. Human currencies are human reference values. Most are now fiat currency based only on a government's promise to pay. Thermodynamic analyses—and many other process measurements—seem alien to people living by currency-based transactions. Odum sometimes rescaled emergy in dollars that he called "em-dollars."[36] Rescaling does not change thermodynamic ratios if all multipliers are the same currency, but changing value over time because of either price changes or currency inflation can be deceptive. An example is burning cheap energy to get expensive energy. A BTU measure does not change over time.

Basing a transactional currency on a unit of energy might be no worse than basing it on gold. However, this would not be magic because human valuations of anything, gold included, are fickle, varying depending on circumstances. At best, a simple price-cost system only says what buyers would prefer and sellers can afford within the transactional system. But when it is important to safeguard nature, it's imperative to trust evaluations made using nature's currency rather than transactional trade-offs expressed in human currencies.

Energy also helps us design intangibles like information, because everything, including information, has an energy history. Energy transformed it into its present state, just like a tree is a solar battery, and paper is usually a tree transformed by a high-energy process. Information media alone have an energy history. Server farms in the

United States use a little over 1 percent of all generated electricity, and personal computing equipment surely uses several times that.[37] Considerable work is going into reducing the energy requirements for server farms.[38] Work to clean up the paper-making process is under way, but going slower.[39]

However, the information captured in these media has an energy history, too—the energy used to gather and compile it, not all of which is human energy. We would like a high personal-energy return from information systems. If we have to work hard to find and interpret information, we may take the easy route: get along without it and guess. An elegantly designed information system for work is low energy to use both for input and output—low energy both to humans and to energy-fed processes, and built into how work is done, as symbiotic with it as possible. The waste in human information systems is apparent every time one has to "straighten out" an incorrect order or incorrect bill. Just knowing IT cost as a percentage of total cost does not tell anyone very much.

IMPLICATIONS OF A PROCESS THEORY OF VALUE

A business system feasting on high energy yields can indulge in decision rules that waste it. But if the yield from global energy sources is dropping, its basic assumptions start to unravel. The macroeconomic implications of this cannot be foreseen in detail, but a few major points to think about are discussed in the following sections.

Thermodynamic Efficiency

Equating making money to efficiency created much of our present trouble. That foolishness let Toyota outsmart a lot of complacent companies under expansionary conditions. Now expansionary business thinking confounds dealing with a much more serious situation.

Financing Missions

From macro to micro levels, financing missions to cope with Compression is very different from expansionary finance. To finance vigorous learning organizations, financiers must learn to understand missions and must finance learning to *do* more effectively, not input–output, black-box growth. At the macro level (central banks), money that was collateralized from nothing for expansion can be deflated into a black hole and reconstituted for Compression—with appropriate mystique, of course. And government investment has to finance better-quality countermeasures for Compression, not keep the expansion party going, pumping up growth just so businesses can have profit and people can have jobs.

Rapid Adaptability

Vigorous learning organizations must develop people for rapid change and rapid learning, but that is not enough. Financiers must carry them through their learning phase buildups, and redirect their surplus cash to other programs. That's financing of missions, not financing of growth, and it implies a different concept of debt and equity instruments.

Allocation of Commodities or Permits

When no more can be had at any price, a classic market no longer exists. The same is true when aggregate waste has to be capped. No rationing method works unless a critical mass of people understands that it is necessary and can't be bypassed. Systems must be designed to rein in corrupt administration (favoritism), black markets, and counterfeiting. Most systems are variants of per capita allocation, with recipients being able to trade their tickets to consume, sometimes called a cap-and-trade system.

Flat Age Demographic Profile

If we are successful at improving the quality of life, many more people will live into their 80s or longer. If population growth is to stay in check, most economies will have a flat demographic age profile, not the steep age pyramid of yesteryear. (Japan, China, and most European countries already do, but don't yet see it as the new normal.) Many social customs, financial systems, and tax laws are hand-me-downs assuming only that only a small fraction of people are too old to work, and that only a small fraction of youth need extensive preparation for life. To head off social conflict, socially accepted definitions of working age, work responsibility, and social obligation have to change. Everything from family planning to retirement plans will be affected.

Employment and Unemployment

Doing more with less in a Compression economy sounds like a prescription for unemployment. Pressure to keep feeding consumption will be intense just to give people something to do, and which they already know how to do. The United States has been a service economy for a long time, and if we are short of wherewithal, even a flea-market economy distributes what there is to buy food and living essentials—if they can be had.

However, in Compression our work will be creating a high-quality, learning service economy to improve quality of life (it isn't what you have; it's what you can do with what you have). An economy crimped by decreasing energy yields will use more labor (to wring more out of less capital and energy) to improve quality, and will create and use less stuff, but much better stuff. For instance, agriculture would use land more intensively, waste less energy and water, grow better quality food, and probably need more labor to do it. Even heavy industries may produce better quality by regarding themselves as service businesses with a mission, rather than maximizing output. Put people to work fixing our social maladies, like caring for the incapacitated, or altering the behavior of irresponsible psychopaths, substance abusers, or folk otherwise inclined to take but not give—not reinforcing it. Working smarter, not harder and harder, is a necessity.

Little Distinction between Profit, Nonprofit, and Government Work Organizations

Our most critical working organizations need to shift toward multi-stakeholder allegiance, blurring distinctions between for-profit, nonprofit, and governmental organizations. That affects tax codes, regulatory systems, and much else, but we have to end the perception that work done "at the expense of private taxpayers and businesses" is optional. All crucial working organizations have to become truly responsible to society—but operate as fast-adapting networks and information systems, not by complex legislation and bureaucracy.

An End to Hot Wars Using Resource-Intensive Weaponry

Mass destruction is the opposite of environmental sustainability. As long as testosterone contests substitute for dialogue, world peace is elusive—but this gorilla of all wicked problems is impossible to budge without effective world policing that begins to suck the tribal glory out of winning. We might—just might—be able to cool that urge if we can edge much of the global population toward a constitution for learning.

We can't keep doing what we've been doing, but become fixated on a hard transition. A global work constitution should bit-by-bit forge more agreement on wicked problems, thereby increasing the domains in which scientific problem solving can function. However, many people will not understand this until they see it in action, so pioneer work organizations have to start this revolution.

AUDITING VIGOROUS LEARNING

Legal systems assume that people and businesses operating in self-interest require audits. Nonprofits and government agencies need financial audits, too. These can't disappear, but they can simplify. For example, auditors must excavate labyrinths of imaginary complexity, when unearthing financial transactions made for tax reasons.

But how can we check whether a learning enterprise is vigorously pursuing a social mission? By operational audits based on today's examinations for certification, inspector general reviews, or awards like the Baldrige Award. Audits may even question whether the enterprise's mission is well thought out and remains valid. However, complex, time-consuming reviews like those for accreditation of health care organizations, do not assure continued improvement.[40]

Auditing processes can promote the organization's improvement if they are not also used as compliance audits for minimal standards. Those induce fear of sanctions, not learning. Furthermore, passing a minimum requirement is not great performance. Auditees can pressure auditors, too, a la Enron with Arthur Andersen. A learning audit appended to a vigorous learning organization's public report certifies that it does what it says and has the capabilities it proclaims. And the auditor releases the final edit of the report to all stakeholders, not just the audit committee of the board. If both auditors and auditees expect performance quality to improve, audits become learning experiences for both. Performance should improve over the last time, no matter how good the last time was; any degradation is a red flag. Once both parties agree that the purpose of a work organization is performance and learning, not maximizing money return, they have signed on to the constitution for learning.

Qualifications to be an auditor include experience both broad and deep in all three domains of thought in Figure 6.2. Auditors need not be hoary with age, but they should have work organization experience in addition to book learning and should have honed a life skill in sufficient depth to comprehend issues from engineering to ecology. The people being audited should be able to regard auditors as peers who can engage in dialogue on how performance and learning processes can be improved. (People can and do perform roles like this now.)

Of course, auditors need audit training before they begin, and must understand what a vigorous learning organization should be. To prevent undue coziness, auditors should be paid by the public or a third party and report to society. Such auditing is a public service profession, not a profit-maximizing opportunity (sometimes forgotten by CPA firms). That is, the auditors have signed on to the learning constitution and have learned to live by it.

Audits begin with the organization's mission and the attitude of leadership (and maybe those of the auditors) toward it, gradually working into detailed process examinations, including observing individuals working and checking computer codes. Audit reviews should culminate in the equivalent of highly qualified counseling. Criteria for such audits can be drawn from frameworks like the Baldrige National Quality Award Criteria (which cover educational and governmental organizations, not just business). More can come from safety audits, fire inspections, the Leadership in Energy and Environmental Design (LEED) certification for green buildings, or MBDC's Cradle to Cradle[SM] certification for design for the environment.[41] While such audit criteria do not now exist, enough has been done to make developing them feasible.

Because organized human learning is the heart of a vigorous learning enterprise, audits need to concentrate on it. This may seem mushy, but experienced auditors free to roam can quickly appraise whether or not an organization is really doing what it claims. Evidence can be physically seen, data can be cross-checked, and watching people at work reveals whether they are putting on a show. But as with all audits, the system must prevent undue influence on the auditors. No system is noncorruptible.

COMMITTING TO THE LEARNING CONSTITUTION

Very little is written in this book that others have not said better, and it never can be complete. It is barely a starting point, for vigorous learning is a process, not a doctrinaire plan. For such momentous change, no blueprint is possible, no step-by-step manual, no sure-fire leadership guide. Every situation is different because every organization has a different cast of characters with its own history. Almost everything cited has been done by real people somewhere, so putting it all together is not impossible; but to have vigorous learning, much has to offset the human instinct to abort it. Notably, the haves who benefit from any existing system have to be prevented from finagling to continue it.

To develop vigorous learning organizations, development of logical system structures and behavioral shifts have to progress together. Even strongly motivated humans can't make such a profound shift quickly, but we can't wait on a new generation either, for Compression calls for millions of people to very quickly embrace work that is very different from that of expansionism. We who are alive and active must commit to rapid change.

Vigorous learning is so different that work organizations that commit to it must free themselves to try totally new approaches partially isolated from expansionist influences, not as an offshoot of economic development programs promoting expansion. When everything is open to change, more imagination is needed.

The experiments of postwar Japan, relatively isolated from commercial culture, created modern quality practices and the Toyota Production System, our best-known counterintuitive work system today. Now we need new experiments on a much bigger scale to meet the challenges of Compression. There are only a few sketchy examples to copy.

Present examples of collective learning remain short of what is needed. For example, Vistage International consists of 14,000 company CEOs regularly meeting in small groups for exchanges and mutual advice. Companies collaborate with nonprofit organizations on precompetitive research, and network collaboration results in wiki systems and open system software.[42] And a few noncompeting companies form groups called *consortia* to exchange nitty-gritty improvement practices; the HPM Consortium near Toronto, Canada, is an example. But in all these cases, the purpose of learning is to improve competitiveness in expansion, not to take on the greater collective challenges of Compression.[43]

Environmentalists praise a community effort to deal with environmental issues by industrial symbiosis in Kalundborg, Denmark. Most members are heavy industries and community utilities that use each others' by-products, including heat, practical in a small geographic area. However, all industrial members focus on market competition, so it too is only one more learning stage on the way to coping with Compression.[44]

Local groups are necessary to deal with Compression locally, while a tightly knit supply chain would lend itself to vigorous enterprise globally if members can create trust. A single core learning organization may belong to several enterprises: one organized around a social mission, like developing "diseasome" research into practical preventive medicine; one around product or service life cycles, as with medical equipment; and of course, one dedicated to local environmental improvement.

But first, just *one* existing work organization has to convert to a full-stakeholder organization, pledging to become a vigorous learning organization. That is not trivial. One organization might take the plunge if others signal that they will not harpoon it in the water. These pioneers have to redefine their mission, what work is, what organizations are and do, and therefore how a collective enterprise functions. And to do it, they have to overcome their "tribal suspicions." Wartime heroics are comparatively easy.

Because this is so difficult and so different, "learning constitutions" seem necessary. We might even daily renew allegiance to them with some seemingly corny pledge, such as, "I swear to develop myself, respect others, and work for the benefit of all humanity." After all, we need nothing less than millions of lifetime commitments.

ENDNOTES

Extended version of endnotes available at http://www.productivitypress.com/compression/footnotes.pdf.

1. The Hunger Project site is http://www.thp.org.
2. The World Transhumanist Association: http://www.transhumanism.org.
3. Kurzweil's futurism is at www.kurzweilai.net. He does not lack critics; for instance, Glenn Zorpette, "Waiting for the Rapture," in *IEEE Spectrum*, June 2008 (part of a special report on the topic).
4. George Lakoff, *The Political Mind*, Penguin Group, New York, 2008.

5. Paul Collier, *The Bottom Billion*, Oxford University Press, New York, 2008.
6. Neocorporatism features regular negotiations between private companies, labor, and the government, as in Sweden, Finland, and Norway. Mondragón is the biggest cooperative enterprise in the world with multiple enterprises, all worker owned. Its site is www.mcc.es/ing/index.asp.
7. Wiegand and Watkins are unusual in coming to public attention. Very few whistle-blowers do.
8. C. Fred Alford, *Whistleblowers: Broken Lives and Organizational Power*, Cornell University Press, Ithaca, NY, 2001, pp. 22–25.
9. Integrity International (http://www.soeken.lawsonline.net) and Blowing-the-Whistle (http://www.whistleblowing.org) are the two best-known support organizations.
10. Mauricio Delgado, "To Trust or Not to Trust: Ask Oxytocin," *Scientific American* (Mind Matters), July 15, 2008.
11. Adam Smith, *An Inquiry into the Nature and Causes of The Wealth of Nations*, 1776, Book 1, Chapter 10.
12. Meetings of the American Society of Agronomy are becoming much more integrative; meetings are held jointly with other societies covering fields such as geophysics, climate change, and biodiversity. See http://www.agronomy.org.
13. "Market theory of value" is a more popular phrase for what economists call a subjective theory of value.
14. The labor theory of value did not originate from Karl Marx. David Ricardo proposed it, but probably did not originate it.
15. Because occupations are difficult to categorize, historical comparisons from national statistics have to be overwhelming to suggest much, and there are many caveats.
16. These assumptions are used to simplify exchange models; few economists literally believe them.
17. Martha Stout, *The Sociopath Next Door*, Broadway Books (Random House), New York, 2005.
18. By 2002, behavioral economics had entered mainstream economic modeling, and Daniel Kahneman and Vernon Smith won a Nobel Prize for many contributions to it.
19. The theory has had critics armed with data ever since Eugene Fama developed it circa 1970.
20. Robert Costanza et al., "The Value of the World's Ecosystem Services and Natural Capital," *Nature*, May 15, 1997.
21. Haley Stein, "Intellectual Property and Genetically Modified Seeds: The United States, Trade, and the Developing World," *Northwestern Journal of Technology and Intellectual Property*, Spring 2005.
22. Eric D. Beinhocker, *The Origin of Wealth*, Harvard Business School Press, Boston, 2006.
23. From the CorporateRegister.com site it is easy to navigate to the annual CSR winners.
24. Sekisui CSR were found at http://www.sekisuichemical.com/csr/report/index.html.
25. "Just Good Business," *The Economist*, Jan. 17, 2008, Special Reports Section.
26. This discussion does not belabor the buildup to the big financial bubble and its aftermath.
27. Vaclav Smil summarizes the situation in Japan at http://www.japanfocus.org/-Vaclav-Smil/2394.

28. Physicists claiming that the whole universe is interconnected include, notably David Bohm, *Wholeness and the Implicate Order*, Routledge, New York, 1980; and Seth Lloyd, *Programming the Universe*, Alfred A. Knopf, New York, 2006.
29. Information on PCBs can be found at many credible Internet sites.
30. Jeff Dukes, "Bad Mileage: 98 Tons of Plants per Gallon," Press Release, University of Utah, Oct. 27, 2003.
31. A U.S. DOE site shows the product yields from refining petroleum and big variations between refinery settings around the world: http://www1.eere.energy.gov/vehiclesandfuels/facts/favorites/fcvt_fotw214.html.
32. See http://www.epa.gov/hg for the current status of all mercury hazards.
33. Just one of many such sites is http://www.greenhomebuilding.com/recyclematerials.htm.
34. "End-of-Life Vehicles Directive" Policy Department: Economic and Scientific Study, European Parliament, 2007.
35. "Ecological footprint" to describe sustainable areas originated with Williams Rees and Mathis Wackernagel, *Our Ecological Footprint*, New Society Publishers, Gabriola Island, BC, Canada, 1995.
36. Odum calculated an "em-dollar" by dividing any country's GNP by its total annual energy usage. No matter which currency is used, that only changes the scale of any energy measurement.
37. Estimates by Jonathon G. Koomey, "Estimating Total Power Consumption by Servers in the U.S. and the World," Final Report Feb. 15, 2007 (Lawrence Berkeley National Laboratory and Stanford University).
38. Progress is being made on this, as reported in "Data Center Energy Report: Final Report," July 29, 2008, found at https://microsite.accenture.com/svlgreport/Documents/pdf/SVLG_Report.pdf, now taken down.
39. For example, the Pulp and Paper Technical Association of Canada suggests that energy reductions of up to 80 percent can be had in some subprocesses for making paper, but progress is slow.
40. A review of the The Joint Commission site (http://www.jointcommission.org) will convince anyone of the reach and complexity of health care accreditation.
41. MBDC (McDonough Braungart Design Chemistry) certification is at http://www.c2c certified.com.
42. Satish Nambisan and Mohanbir Sawhney, *The Global Brain*, Wharton School Publishing, Philadelphia, PA, 2007.
43. "Consortia" operate by a semiformula. Top management of each member company site has to be engaged and constitute the governance of it, sustaining the funding for it.
44. You can learn more about Kalundborg at http://www.symbiosis.dk.

About the Author

Robert W. Hall, PhD, is the professor emeritus of operations management at the Kelley School of Business at Indiana University, in Plainfield, Indiana. Dr. Hall is a founding member of the Association for Manufacturing Excellence (AME) and served as editor-in-chief of AME's publication, *Target.* He authored and coauthored six books on manufacturing excellence; the best-known is *Zero Inventories* (1983), one of the first books on what is now called "lean manufacturing." Dr. Hall has provided insight regarding enterprise excellence for over thirty years and has long studied how work organizations might transform to meet twenty-first century challenges. He has been an examiner for the U. S. Malcolm Baldrige National Quality Award and is now a judge for the Pace Award (innovations by auto suppliers). He has reviewed applicants for Industry Week's annual 10 Best Plants competition since 1990. Before academia, Dr. Hall worked for Eli Lilly and Union Carbide. His undergraduate degree is in chemical engineering; his graduate degree is in operations management.

Index